WHEN HUMANS NEARLY VANISHED

WHEN HUMANS NEARLY VANISHED

THE CATASTROPHIC EXPLOSION OF THE TOBA VOLCANO

Riverhead Free Library
330 Court Street
Riverhead NY 11901

Donald R. Prothero

SMITHSONIAN BOOKS

WASHINGTON, DC

This book may be purchased for educational, business, or sales promotional use. For information, please write Special Markets Department, Smithsonian Books, P.O. Box 37012, MRC 513, Washington, DC 20013

Published by Smithsonian Books
Director: Carolyn Gleason
Creative Director: Jody Billert
Managing Editor: Christina Wiginton
Project Editor: Laura Harger
Editorial Assistant: Jaime Schwender
Edited by Juliana Froggatt
Typeset by Scribe Inc.
Indexed by the author
Graphics by the author and Bill Nelson

Library of Congress Cataloging-in-Publication Data

Names: Prothero, Donald R., author.
Title: When humans nearly vanished : the catastrophic explosion of the Toba volcano / Donald R. Prothero.
Description: Washington, D.C. : Smithsonian Books, [2018] | Includes bibliographical references and index.
Identifiers: LCCN 2018008391 (print) | LCCN 2018013582 (ebook) | ISBN 9781588346353 (hardcover) | ISBN 9781588346360 (ebook)
Subjects: LCSH: Supervolcanoes. | Volcanoes. | Survival.
Classification: LCC QE522 (ebook) | LCC QE522 .P76 2018 (print) | DDC 551.2109598/12—dc23
LC record available at https://lccn.loc.gov/2018008391

Manufactured in the United States of America
22 21 20 19 18   5 4 3 2 1

This book is dedicated to my wonderful wife,

Dr. Teresa LeVelle,

and to our wonderful friend

Kim Milliken Hayden

for all her support

Civilization exists by geologic consent, subject to change without notice.

Will Durant, "What Is Civilization?," 1946

CONTENTS

PROLOGUE

A Very Bad Day on Planet Earth

Our story begins on a very bad day about 74,000 years ago. The planet was starting to move out of one of its more recent ice ages, although in the tropics there was little change in climate between the Ice Age glacial episodes and the warmer interglacial episodes. A wide range of late Ice Age mammals inhabited the world, including woolly rhinoceroses and mammoths up in the cold regions of Eurasia, along with huge bison, giant deer, wild horses, and a variety of smaller mammals. Giant lions, saber-toothed cats, and huge bears fed on these large prey animals.

People lived in many parts of the Old World by then but had not yet reached Australia or the Americas. The bulk of the human population were archaic members of our species, *Homo sapiens*, which first appeared in southern Africa about 100,000 to 300,000 years ago. By 74,000 years ago, these people had spread out of Africa and may have occupied much of Asia, as well as parts of southeastern Europe. However, Europe was still dominated by another human species, the Neanderthals, who had adapted to life on the edge of the northern ice sheet. In contrast to archaic *Homo sapiens*, Neanderthals had a shorter, stockier, more muscular build and shorter limbs, a body type suitable for attacking large prey and adapted for reducing heat loss in the cold climate. In the far reaches of Asia, ancient humans had spread to many of the islands of modern-day Indonesia and Malaysia. On one now known as Flores, east of Java and Bali, they evolved into a dwarfed species, *Homo floresiensis*. Now nicknamed "hobbits," these people stood only about 1.1 meters (3 ft, 7 in) tall, shorter than any modern adult Pygmies (given their small brain size, some anthropologists question whether they are even in our genus, *Homo*). Flores is part of the island chain (including Sumatra, Java, and many smaller islands of the Malay Archipelago) that makes up of most of modern Indonesia. These islands are built completely of volcanoes, both active ones and ancient, dormant ones. Their climate is tropical and their jungle is dense. So much vegetation grows on the rich volcanic soil, in fact, that it's often hard to recognize signs of volcanoes there.

About 2,000 kilometers (1,200 mi) northwest of Flores, in northern Sumatra, there are numerous volcanoes that have erupted over the past million years. On the very bad day in question, one particular volcano, now known as Mount Toba, had been active for a long time. It had bulged up gradually until it towered almost 900 meters (3,000 ft) above the jungle. All around this monstrous dome, cracks formed. Hot springs and fumaroles spewed out steam that smelled like rotten eggs because of all the sulfur in the mixture. Earthquakes both large and small had rocked the entire island of Sumatra for a year before Mount Toba started to blast out a few small eruptions of steam and ash that blanketed the surrounding jungle. These were likely terrifying to the local animals and people but soon forgotten once they had quieted down. After a few years of tropical rains and rapidly growing jungle, the blanket of ash had vanished. Yet recently the frequency of smaller eruptions venting steam and ash had begun to increase. Soon the sides of the volcano turned barren and rocky as red-hot ash and balls of scalding pumice burned away all the nearby jungle.

This was the state of things as that fatal day about 74,000 years ago dawned. Events really got rolling when the huge dome rumbled with a deep vibration that shook all of Sumatra. Jets of steam and ash shot one after another from the summit. Then came an explosion louder than any sound previously heard by humans in their entire evolution. For comparison, when the Krakatau (or Krakatoa) volcano, also in Indonesia, erupted in 1883, it created a sonic boom that could be heard 8,000 kilometers (5,000 mi) away and which traveled around the world seven times. That blast, 5,000 times as powerful as the Hiroshima nuclear bomb explosion, was the largest explosion heard in recent times. Yet the eruption of Mount

Toba released the energy of a million tons of explosives, 40 times larger than the largest hydrogen bomb that humans have ever built, more than 1,000 times as powerful as Krakatau, and 3,000 times as powerful as the eruption of Mount Saint Helens in 1980. Thus, the sonic boom from Toba must have been deafening to animals and people for many kilometers around and must have bounced around the earth repeatedly, dwarfing any other sound produced on earth in the previous 28 million years.

After the explosion, a gigantic mushroom cloud of hot ash rose many thousands of meters into the stratosphere. Meanwhile, superheated ash and gases, as hot as 1,100°C (2,000°F), flowed down the sides of the mountain in gigantic, turbulent clouds traveling up to 320 kilometers (200 mi) per hour. They incinerated everything in the jungles for many kilometers. A dense blanket of ash and pumice covered not only Sumatra but also most of the nearby islands, causing death and devastation wherever it settled. The ashfall spread across southern Asia as well, leaving a thick deposit even in India, more than 3,000 kilometers (1,800 mi) away. The ash cover in India was on average about 15 cm (6 in) thick; in the years following the eruption, it mingled with other layers and moved downslope, forming secondary ash deposits several meters thick (as happened in the Mount Saint Helens eruption in 1980).

Tropical rains turned the ash into mud with the texture of wet cement, which made rivers and pathways into impassable morasses and collapsed tree branches and sometimes even whole trees under its weight. Small huts, too, probably were crushed under thousands of kilograms of wet ash. Sea levels were lower at that time, but it is likely that a tsunami triggered by seismic activity associated with the eruption would have killed many people living along the coast. The people and animals of the jungle found their world in utter ruin, and most local survivors must have soon died of starvation, while others died from inhaling dry, dusty ash. Volcanic ash particles are microscopic shards of glass, and they cut up the insides of the lungs, which scar and then clog with fluid.

These were the effects on the life of the jungles within a few thousand kilometers of the erupting volcano. But areas beyond the densest ashfall were affected as well. Clouds spread around the world, leaving a blanket of ash on the ocean floor in many places thousands of kilometers from the eruption. The volcano spewed out about 10 billion metric tonnes (11 billion short tons) of sulfuric acid and 6 million metric tonnes (6.6 million short tons) of sulfur dioxide, which combined with water in the atmosphere to make sulfuric acid. The sulfuric acid was devastating to life in many parts of the globe and can be detected even in the Greenland ice sheet.

The farthest-reaching impact of the eruption, however, was caused by the 3,000 cubic kilometers (720 mi³) of dust-size particles of volcanic debris that were injected into the stratosphere, more than 10 kilometers (6 mi) above sea level. At that altitude, they were picked up by the jet stream, and soon a plume of ash began to circle the world. When something like this happens, the amount of sunlight that reaches our planet is reduced, resulting in abnormal cooling. When Krakatau erupted in 1883, the huge volume of ash that was shot into the stratosphere blocked sunlight, and global average temperatures dropped by about 1° to 2°C (2° to 4°F) for more than a year. Weather patterns were erratic for years, and temperatures did not return to normal until 1888. The sky was dimmed, even darkened, for months after the eruption, and the large amount of particulate matter in the stratosphere changed its color, producing, for instance, spectacular orange-red sunsets, such as the one depicted in Edvard Munch's *The Scream* (1893). As Munch wrote in his diary on January 22, 1892: "Suddenly the sky turned blood red. . . . I stood there shaking with fear and felt an endless scream passing through nature." Rare atmospheric effects, including a literal blue moon, a Bishop's ring (a faint brown halo around the sun), and volcanic purple light at twilight, were also seen around the world.

Sixty-eight years earlier, when Mount Tambora (also in Indonesia) had erupted in 1815, it injected so much dust into the stratosphere that the earth's weather patterns changed. As the ash blocked sunlight, the resulting cooling led to crop failures, starvation of livestock, and widespread disease (including a typhus epidemic) and famine in human populations around the world. The following "Year without a Summer" (1816) saw cold, dark, rainy summer months in North America and Eurasia: even in June, it snowed in New York, New England, and many European cities. That month, Percy and Mary Shelley were staying at Lord Byron's villa near Lake Geneva in Switzerland, and they told one another gothic horror stories to pass the long hours spent indoors. That wet, gloomy summer inspired Mary Shelley to write *Frankenstein, or The Modern Prometheus*.

The eruption of Toba about 74,000 years ago was 1,000 times as large as that of Tambora or Krakatau. It didn't just trigger a summerless year or a short cold spell spanning several years: global temperatures dropped by 3° to 5°C (5° to 9°F), to a worldwide average of just 15°C (60°F) after three years, and took a full decade to recover to pre-eruption levels. The tree line and the snow line fell to 3,000 meters (10,000 ft) lower than where they are today, making most high elevations uninhabitable. Ice cores from Greenland show the evidence of this dramatic cooling in trapped ash and ancient air bubbles.

What happened to people and animals during this terrible time? As we shall see in the rest of this book, many geneticists and archaeologists believe that the Toba catastrophe nearly wiped out the human race; afterward, they argue, only about 1,000 to 10,000 breeding pairs of people survived worldwide. Supporting this idea are both geologic evidence of Toba's size and atmospheric effects and indications of a human genetic bottleneck that happened around the time of the eruption. A genetic bottleneck occurs when the number of individuals in a population drops so low that its genetic diversity is greatly reduced, and all descendants of that population carry the rare genes of the handful of survivors.

Several studies have found similarly timed bottlenecks in the genes of human lice and of our gut bacterium *Helicobacter pylori*, which causes ulcers; according to these organisms' molecular clocks, which show how much time has passed since a genetic change took place, both bottlenecks date back to the time of Toba. The molecular clocks of a number of other animals, including tigers and pandas, indicate that they, too, passed through a bottleneck around that time. In short, Toba was the biggest eruption since modern humans appeared on earth, and it came very close to wiping out people, along with many other animals, altogether.

The Toba eruption was one of the greatest geological catastrophes ever to strike our planet. It was larger than any volcanic eruption in the previous 28 million years and hundreds to thousands of times larger than later eruptions such as Tambora, Krakatau, and Mount Saint Helens. Toba may even have been a disaster on the scale of the one 65 million years ago that wiped out the dinosaurs and many other creatures, and it may have been similar in its effects to other mass extinction events in our planet's history.

However, the amazing story of the Toba eruption, and its aftereffects, is one that few people (and even few scientists) have heard. Not until the late 1990s did researchers even realize that the catastrophe had occurred; at that point, many scientists, working on many different kinds of problems in geology, genetics, and other fields, eventually came to recognize that they were all uncovering evidence of the same great disaster. The story of Toba's discovery, to which we now turn our attention, is one of surprise, serendipity, and continuing controversy.

Mystery of the Missing Megavolcano

The most exciting phrase to hear in science, the one that heralds new discoveries, is not "Eureka!" but "That's funny."

Isaac Asimov, 1987

Serendipity

Percy Spencer was a remarkable character. Born and raised in rural Maine, he was just a baby when his father died, and he grew up poor. At age 12, he left school to support his family, working from sunup to sundown in a spool mill. Even though he had only a grammar school education, he learned the basics of electrical engineering while helping to bring electricity to a local paper mill and then to his rural community. He first read about wireless communication in news coverage of the *Titanic* disaster and volunteered to join the US Navy so he could learn all about radio technology. While standing watch at night, he read voraciously and taught himself trigonometry, calculus, physics, chemistry, and metallurgy.

By 1939, he was one of the world's leading experts on radar tube design, building radar and improving its technology at Raytheon to aid Allied defenses in World War II. He found a more efficient way to manufacture the magnetrons that generated the microwave energy for radar sets, and soon daily production of the units increased from 17 to 2,600. When the war ended, however, Raytheon was stuck with an assembly line building thousands of military-grade magnetrons and no use for them. Then one day, while standing near a magnetron, Spencer noticed that a candy bar in his lab coat pocket had melted. This had previously happened to others, but he was the first to realize that this accidental discovery might point to a civilian market for all the excess production. Over the next few months, he and others at Raytheon experimented with the magnetrons, and in 1947 they produced the first commercial microwave oven. Even though Spencer was the one who filed the patent for the device, he received no royalties, only a two-dollar gratuity from Raytheon for his invention (he did end up as a senior vice president and a member of the company's board). This amazing invention, which nearly every modern US household depends upon today, was based on an *accidental* discovery.

Then there is the story of Arno Penzias and Robert Wilson. In 1964, they were employed by Bell Labs, the original research division of AT&T/Bell Telephone, and were responsible for developing communication technology for "Ma Bell." They were the chief electrical engineers who designed the first microwave antennas that received and transmitted phone signals, and their main job was to get the bugs out of the devices and improve their efficiency. The antennas were picking up unwanted static and other electronic "noise," and Penzias and Wilson found and resolved most of the interfering sources. Yet there was one source they could not eliminate. The noise was 100 times stronger than they expected. It occurred day and night and was evenly spread across the sky (and thus wasn't coming from any single point source on earth or in space). It was clearly coming from outside our galaxy, and they could not explain it. Luckily, just 60 kilometers (37 mi) away, in Princeton, New Jersey, the physicists Robert Dicke, Jim Peebles, and David Wilkinson were beginning an experiment to find the electromagnetic noise left over from the Big Bang, the explosion that, it is theorized, began the universe. A friend told Penzias that he'd seen a preprint of a paper by the Princeton group predicting this background noise. The Bell engineers got in touch with the Princeton lab, and Penzias and Wilson showed the physicists what they had found. Lo and behold, the two Bell Lab engineers had uncovered proof, in the form of cosmic background radiation, that the

Big Bang had happened. In 1978, Penzias and Wilson received the Nobel Prize in Physics—for something discovered entirely by accident.

These stories are classic examples of how "pure" scientific research can lead to amazing findings. Some discoveries are made by people looking for a specific answer to a specific problem, but more often than not, scientists reach important breakthroughs by doing pure research—gathering a broad range of data on a particular topic without knowing what they might find. Many people think that science is about planning your research carefully to achieve some specific goal, and often there is little patience or tolerance for pure research: investigations that don't have a particular conclusion in mind but instead focus on finding out general facts about nature, whether or not those facts might prove to have practical applications. Even many scientific funding agencies show such impatience, tending to reward investigations that are conventional and look for "more of the same." Agencies seldom fund research that is speculative, perceived as a gamble, or seen as exploration for its own sake that is not directed toward a specific practical goal. TV commentators and congressional representatives sometimes ridicule research that has no particular aim or, especially, obvious application; occasionally, even when such research has been approved through the well-established scientific review process, it is shut down.

The irony of the misconception that science must be practical and useful is that most great scientific discoveries are not anticipated or planned but happen by accident. More often than not, scientists who find a crucial new piece of evidence were not actually looking for it but instead were searching for something else, making their great discovery without planning or expecting it. The term *serendipity* was coined to describe such a phenomenon. It comes from an old Persian tale, "Three Princes of Serendip," whose protagonists make discoveries unexpectedly. In the case of science, serendipity is most often seen when a researcher is prepared to recognize the implications of some new, unanticipated development. As Louis Pasteur put it in a lecture at the University of Lille in 1854, "In the field of observation, chance favors only the prepared mind."

Other examples of serendipity in science are legion, especially in chemistry. Alfred Nobel created gelignite, the key ingredient in his development of TNT, by accidentally mixing nitroglycerin and collodion (guncotton). Hans von Pechmann inadvertently discovered polyethylene in 1898. Silly Putty, Teflon, superglue, Scotchgard, and rayon were all chance inventions, as were the discoveries of the elements helium and iodine. Among drugs, penicillin, laughing gas, the birth-control pill, and LSD were all based on chance discoveries, too. Viagra was originally developed to treat blood pressure, not impotence, and minoxidil to treat ulcers, not hair loss. Among other practical inventions, ink-jet printers, cornflakes, safety glass, Corning Ware, and the vulcanization of rubber all owe their genesis to happy accidents. Many great advances in physics and astronomy were also unexpected, including the discoveries of the planet Uranus, infrared radiation, superconductivity, electromagnetism, and X-rays.

Likewise, geologists often find things they are not looking for. In 1855, John Pratt and George B. Airy were doing routine surveying for the British government in northern India. They noticed that the plumb line under their surveying tripod was not as gravitationally attracted to the Himalayas as they expected it would be. This led them to discover evidence for the deep crustal roots of such mountains.

Similar examples could be cited for many more pages, but the point is clear: science is not always predictable, and scientific research cannot be restricted to looking for only straightforward results that were expected when the study began. Otherwise, important discoveries will come to an end.

Frozen in the Ice

The discovery of the Toba eruption is another classic case of scientific serendipity. In the late 1980s, several researchers working in labs around the world and with entirely different goals turned up indications of the catastrophe completely by accident. Most of these researchers were looking for evidence not of a supervolcano but of how local and worldwide climate had changed during the most recent Ice Age. That was, and remains, an extremely important question in and of itself, and it has

Figure 1.1. Drilling an ice core at the EastGRIP site in Greenland, 2017. (Photograph by Helle Astrid Kjaer; courtesy Wikimedia Commons)

led to even more exciting discoveries as scientists have systematically combed through ancient climate records—captured in ice cores, in cores of sediment drilled from the deep seafloor, and in deposits of ancient volcanic ash found both on land and in the world's lakes and oceans—to see what they might reveal.

Ice cores are a good example of how serendipity can lead to accidental discoveries when scientists conduct pure research without knowing what they might find. Labs in countries all over the world have been drilling cores in many spots in the Antarctic and Greenland ice sheets, as well as in larger glaciers on mountain ranges, for several decades (fig. 1.1). Because bubbles captured in this ancient ice consist of tiny samples of air from thousands and even a million years in the past, these cores have given us crucial insights into the climate puzzle of the ice ages.

One of the first scientists to use the chemistry of ice cores in this way was Willi Dansgaard, a Danish climatologist who took advantage of US military spending at the height of the Cold War to do his research. In the 1950s, the US Army decided to build a secret military base under the ice of northwestern Greenland to test the feasibility of constructing more such bases, which could be used to both monitor Soviet activity in the Arctic and hide up to 600 nuclear missiles a short hop from the Soviet Union. For this program, dubbed Project Iceworm, the Army Corps of Engineers built a network of long ice tunnels, which was named Camp Century and was intended to spread out over some 4,000 kilometers (2,500 mi) about 8.5 meters (28 ft) below the ice surface. It was powered by its own nuclear reactor and housed as many as 200 personnel during its peak years of activity in the early 1960s.

Some of these personnel were nonmilitary scientists, who were using the army's dime to pay for their Arctic research in return for helping it to understand risks and problems associated with the base. Soon the scientists gave the military engineers bad news: glaciers are rivers of ice that constantly flow from higher regions to lower ones, and the ice sheet in which Camp Century had been tunneled was flowing so rapidly that it would soon crush the reinforced bunkers beneath the ice. Within three years, the predictions came true, and in 1966 the military had to abandon the base. It was completely flattened under the ice, which global warming is now rapidly melting away.

Dansgaard was among the civilian scientists on the base. When he arrived at Camp Century in 1964, he was already a 42-year-old, bearded, pipe-smoking, grizzled veteran of many polar expeditions, including several voyages to measure the chemistry of the ice and the seawater around Greenland. His wife, Inge, was also a polar veteran. She had been sent to Greenland in 1947, first to study the earth's magnetic field and later to study meteorology, with Willi tagging along. He showed a knack for devising simple, minimally expensive solutions to scientific problems: for example, no one before the 1940s had known exactly what the chemical composition of rainwater might be, so Dansgaard placed a funnel in the neck of a beer bottle in his yard and collected a clean sample to analyze.

He received permission from the military brass to study ice cores that the military and civilians had drilled out of the ice sheet next to Camp Century from 1961 to 1963 to assess the ice sheet's stability. Eventually, these pioneering cores, taken all together, spanned more than 1,300 meters (4,300 ft) and covered 100,000 years of time. They yielded the first evidence of the operation of both glacial-interglacial cycles and the El Niño–Southern Oscillation (ENSO) climate cycle, but perhaps the most startling scientific findings of the Camp Century cores was that ice ages lasting almost 100,000 years could end abruptly, in only a few thousand years. In collaboration with the Swiss geochemist Hans Oeschger, Dansgaard discovered the rapid temperature fluctuations, now known to all climatologists as Dansgaard-Oeschger cycles (D-O cycles for short), that occurred every few thousand years during the last glacial period, which ran from about 110,000 to 11,700 years ago. Thus, although the US Army's attempt to build a polar base proved to be a failure, it serendipitously led to one of the great scientific breakthroughs of the past century.

More recently, in 2004, EPICA (the European Project for Ice Coring in Antarctica) published its results from cores drilled in the Antarctic ice sheet that span the past 680,000 years of the earth's climate and record at least seven complete glacial-interglacial cycles. These cores have been crucial in proving that modern global warming is human-caused, not natural: during even the warmest interglacial episodes of the past 680,000 years, the atmosphere contained no more than 280 parts per million (ppm) of carbon dioxide, while carbon dioxide concentration in the atmosphere during our current warming episode has shot far past that threshold. In 2018, we passed 410 ppm, a concentration higher than that of any time in the past 50 million years.

Another significant ice core was drilled in 1993 by a team of American, Swiss, and Danish researchers as part of an international research effort named Greenland Ice Sheet Project 2 (GISP2), whose first phase had lasted from 1971 to 1981. Taken from the highest part of the Greenland ice pack, the GISP2 core set a length record: it was more than 3,000 meters (10,000 ft) long. (The EPICA core, at 3,233 meters [10,600 ft], has since surpassed it.) Each segment of the core was only about 3 meters (10 ft) long, as it was divided for easier handling. After each piece emerged from the drill hole, researchers stored it in a deep freezer, later bisecting it for study and then wrapping it up and replacing it in the freezer. After undergoing preliminary study at the drill site, the core segments were shipped to researchers via refrigerated airplane; today most are stored at the National Ice Core Laboratory in Lakewood, Colorado.

Among the scientists who examined the GISP2 core was Greg Zielinski, then at the Climate Change Research Center of the University of New Hampshire. While carefully sampling and analyzing the chemistry of air bubbles throughout the core, he was particularly struck by those dated about 74,000 years ago: their concentration of sulfur dioxide, a component of sulfuric acid, was way off the scale. His analysis suggested that there were 2,000 to 4,000 megatons of sulfuric acid in the atmosphere then, 25 times more than all the industrial sulfuric-acid pollution produced over the entire earth today. He knew that this concentration of sulfuric acid in the atmosphere must mark some extraordinary event. Only a volcano could have produced so much so quickly, but when Zielinski first published this information, in 1994 (just a year after the GISP2 core had been drilled), he could not pin down the source of the sulfuric acid.

Data in Davy Jones's Locker

Another classic example of how great discoveries happen by accident, or simply as a result of collecting a lot of data, is seen in marine geology and the contributions it has made to climate studies. During World War II, the US government poured money into a huge number of research labs that were working on inventions that might prove to have military use; eventually these labs would develop radar, invent the atomic bomb, and make numerous improvements to gun technology, tanks, aviation, and ship building and design. (My father contributed to some of these improvements: at Lockheed his entire career, he had a hand in the development of every plane the company built, from the P-38 Lightning fighter of World War II fame to jet fighters to legendary supersecret spy planes such as the U-2, SR-71 Blackbird, and Stealth fighter.) When World War II ended, the government kept some of the big labs going, both to support the military during the Cold War and for civilian research purposes. The continued push for research led to the United States' development of the hydrogen bomb, the B-52 bomber, satellites, rockets, and many other important weapons systems. Some labs also received government support for nonmilitary projects that might yield practical benefits for the country.

Among these projects was research into the oceans and marine geology, a need already apparent during World War II, when many lives were lost on surface ships and in submarines because almost nothing was known about the seafloor. Maps and charts were poor, and key aspects of ocean currents were misunderstood. In fact, before World War II, we had no idea what over 70 percent of the earth's surface looked like because there were no detailed maps of the ocean floor. By the late 1940s, several oceanographic institutes were beginning to fill this huge gap in our knowledge of the earth's surface. They included Woods Hole in Massachusetts, Scripps in San Diego, and the Lamont Geological Observatory (now Lamont-Doherty Earth Observatory) of Columbia University in New York, just an hour up the Hudson Valley from New York City.

Lamont's first director was William Maurice Ewing (known to everyone as Doc), a driven, irascible Texan geophysicist trained at Rice University in Houston. Along with Jack Nafe (who taught me marine geophysics when I was at Lamont), he pioneered the technique of using dynamite explosions to bounce sound waves off the sea bottom, analyzing those reflections to "see" the structure of the layers below. Already a professor of geology at Columbia, in 1949 Ewing became the director of the new institution. It was housed on the weekend estate of the Corliss Lamont family, who had donated the land to Columbia. As the science writer William Wertenbaker described Ewing in his book *The Floor of the Sea* (2000):

> While he was an exacting scientist with excellent instincts, above all he had a passion to know. It was as explorer of two-thirds of Earth's crust—directly and through the generations of students he inspired—that he had his greatest and most lasting influence. Before science could explain it, Ewing saw, perhaps better than anyone, how much there was to learn about the seafloor. He encouraged

and he pushed and he bullied and he insisted on the collection of data (oceanwide at first and worldwide later) at all times—as much data and of as many kinds as men and machines were capable. Maybe more.

Ewing was an amalgam of unblended traits and contradictions: a Texas farm boy dedicated in mind and body to the sea; an inspirational presence sometimes asleep in his own class; and a courteous Southerner who was a terror to the hidebound.

Ewing was not only a brilliant scientist but an ingenious inventor as well, with a knack for simple practical solutions to problems (e.g., when seawater ruined his hydrophones, used for picking up sound vibrations in the oceans, he sealed the next set of hydrophones in over-the-counter rubber sheaths). He designed and developed many of the instruments for oceanographic research that have since become standard. Skilled in many fields of science and extraordinarily hardworking, he didn't take well to people who were less industrious, according to Wertenbaker:

Many a brilliant man has not fulfilled his promise. Ewing had brilliance, but he had more. At Rice, his labs and classes were from nine in the morning to four in the afternoon. From five in the afternoon to midnight he held jobs to help pay his way. And there was also the need to fit in homework, meals, and time with friends.

The resolution for maintaining this schedule was pure Ewing. "In some of my freshman reading, it said that sleep is deepest and most restful in the first two hours. I thought about that. It looked to me that I could just sleep for two hours twice a day and then have time to do all the stuff I had to do."

It worked pretty well for a while. Then a professor invited Ewing to dinner. Before the meal, his host poured wine for his wife, [his] daughter, Ewing, and himself. Ewing had never so much as seen wine before. He was on his best behavior, sitting chatting at one end of a long sofa. The next thing he knew, he awoke in a bed upstairs and it was daylight.

Years later, sleep still took second place to science. "Once, Doc joined the ship at Nassau," said colleague Charles Drake. "He looked exhausted. He probably was. He usually is. I took the first watch, he took the second. He looked so pooped that I let him sleep that night, and stood his watch after mine. Along about 4 a.m. up comes this real mad body. 'Why didn't you wake me up?' he asked. 'Well, you looked tired,' I said. 'Don't you ever do that again!' he yelled. And bow-wow-wow-wow-wow."

By 1953, Lamont had its first ship, the R/V (research vessel) *Vema*. Rather than use it for one specific mission, Ewing assigned it to cruise the entire globe, collecting every bit of data imaginable. In 1962, the R/V *Conrad* joined the Lamont fleet. At their peak, these two ships traveled 64,000 kilometers (40,000 mi) annually, staying at sea almost 300 days a year. Ewing had them record the depth of the ocean floor wherever they went, take water samples at many depths for temperature and salinity, measure the strength and temperature of ocean currents, collect data on the magnetic patterns of the seafloor, and bounce sound waves off sediments on the bottom to understand their structure. His depth soundings led to the amazing ocean floor maps (representing 70 percent of the earth's surface) prepared by the Lamont scientists Marie Tharp and Bruce Heezen, which have since become iconic. Most important were Lamont's discoveries of features such as the midocean ridges, which traverse entire seas and are longer and higher than any mountain range on land. Such ridges proved crucial to the revolutionary scientific understanding of plate tectonics, which emerged (partially from the work of Lamont scientists) in the 1960s.

In addition, Ewing issued a standing order that each ship would take a deep-sea core at the end of every working day while it was stopped for the night, no matter where it was. Many of these cores turned out to yield crucial evidence on the history of oceans, climates, and the evolution of life. Beginning in the 1950s, oceanographic vessels from many institutions in addition to Lamont were routinely collecting long cores of sediment by dropping a straight piece of steel pipe down into the sea bottom, then

Figure 1.2. A piston corer, a 10-meter- (30-ft-) long steel tube used to recover a long cylinder of oceanic and lake sediments accumulated over a span of millions of years from the seafloor. This corer, seen in 2004, is taking samples of sediment from the bottom of Lake Tanganyika, a crucial site in the argument about how the Toba eruption affected climate. (Courtesy Nyanza Project, University of Arizona)

pulling it back up to the boat (fig. 1.2). Although Danish and British institutes collected some cores, American ones made the most intense effort.

The Lamont core collection is an immense database that has solved a host of scientific questions in climate and earth history. It records an almost continuous story of what has happened in ocean waters over millions of years. The sediments in these cores include not only fine grains of mud eroded from nearby landmasses but also microscopic shells of tiny plankton smaller than a speck of sand. When these plankton die, they rain down from the ocean's surface waters to the seafloor, forming an unbroken record of what happened while they were alive: how warm the ocean waters were, what the ocean chemistry was, how they evolved or migrated in response to climate change, and many other important pieces of information.

I spent much of my career in graduate school examining the shells of plankton in deep-sea cores from all the world's oceans that are stored at Lamont-Doherty Earth Observatory. Like the GISP2 ice cores, they had been carefully sliced in half lengthwise and wrapped up for storage once they reached the lab area on the research ship that collected them. They now occupy the Core Lab at Lamont, a barnlike structure larger than a football field with racks up to the ceiling, which houses more than 20,000 cores that together are more than 72 kilometers (45 mi) long.

The core labs at Lamont, Scripps, and Woods Hole are crucial resources for anyone studying the deep history of oceans and climate. Among the people who sift through this story again and again is Mike Rampino, formerly of the Goddard Institute of Space Science in New York City and now at New York University. I've known him as a colleague since the early 1980s, and he has always been fascinated by possible connections between big events, such as volcanic eruptions, asteroid impacts, climate change, and the evolution of life. In the early 1990s, Rampino happened to be looking at planktonic microfossils in a series of cores drawn from many oceans. The minerals in the shells of these plankton trapped oxygen as they precipitated downward in the seawater, thus "recording" the temperatures of the ocean water in which they lived. (Water that contains oxygen's lighter isotope, oxygen-16, evaporates more easily than water with its heavier isotope, oxygen-18. In normal ice-free conditions in the ocean, oxygen-16 evaporates as part of the water cycle, eventually returning to the ocean in snow and rain. But when large ice sheets lock up lots of oxygen-16-rich water at the poles, ocean waters grow richer in oxygen-18. This geochemical process is a great recorder of ice volume and thus temperature.)

One day Rampino noticed that the oxygen isotopes in his planktonic microfossil samples showed a cooling of 5° to 6°C (more than 10°F) over a period of just a few hundred years. This suggested that a very rapid fluctuation in climate had occurred, even faster than that seen at the beginning of a typical ice age. At the same level of the cores in which this rapid change was seen, there was an ash layer. It had first been noted by the Lamont scientists who described the core back in the 1970s. When Rampino calculated the ash's age, he found that it was about 74,000 years old. At the time, he did not know of Zielinski's discovery of excess sulfur dioxide in ice cores of the same age. But a few months later he stumbled upon Zielinski's work and quickly realized they'd found something important. When two completely different lines of evidence (in this case, ice cores from Greenland and sediment cores from the bottom of the ocean) lead to similar answers, something really big could be going on.

Very few geological phenomena can cool the planet so quickly. One is the impact of bodies from space, such as asteroids or comets. Such impacts are capable of blasting huge amounts of dust into the stratosphere and causing nuclear winter–like conditions. But they don't release huge amounts of sulfuric acid. Only an enormous volcano could do that. But which one? Just one or two of the larger volcanic eruptions on the planet had happened close to 74,000 years ago, but none that had been dated and published so far was of the right age.

Blowing the Case Wide Open

The search for the cause of the mysterious data gleaned from ocean cores and ice cores soon shifted to the realm of volcanologists: scientists who study volcanoes, including the gases that steam out of volcanic vents; microscopic details of slices of volcanic rock; the structure of volcanic features in the field; and the patterns of earthquakes that eruptions produce. One subgroup of these scientists are known as tephrochronologists, from *tephra*, a term describing a mixture of volcanic debris including ash and pumice, and *chronos*, "time" in Greek. Tephrochronologists study and date volcanic ash layers wherever they are found around the world. These layers are tremendously powerful tools in geology. Not only do they tell us when a particular volcano erupted and how far its ash spread, but they also serve as time markers that allow us to correlate events across the globe with high precision. If we find the same ash in a deposit on land and in a deep-sea core, we can be certain that it represents the exact same moment in these two completely different realms. Then we can use this marker to assign a relative chronology to other events that cannot be dated as precisely.

A handful of the very best tephrochronologists can pin down the age and source of almost any volcanic ash layer through detailed study of its particles and especially its unique geochemical signature. Among them is John Westgate of the University of Toronto, who has sampled and studied hundreds of different ashes from around the world. In 1990, scientists across the globe began sending him samples of a mystery ash that, oddly, all seemed to have come from the same source, although they were found in a swath more than 6,500 kilometers (4,000 mi) wide. They were evidence of a huge eruption that would have been more powerful than any in recorded history.

The best way to confirm that ash samples are all from the same volcano is to date them. When you look at tiny shards of volcanic glass under a microscope, you can see marks caused by the decay of the rare isotope uranium-238. Trapped inside the original magma and decaying by nuclear fission, the escaping nuclear particles disrupt the structure of the glass samples and leave fission tracks. These tracks accumulate over time, so by counting the tracks, you can date the sample. Westgate sifted the glassy particles out of the ash samples with magnetic separators. When he analyzed them, all the samples had the same age: 73,000 ± 5,000 years. Therefore, they were all from the same volcano, which must have produced an ash cloud that went farther around the world than any other ever known.

Yet no one had published a record of a volcano with the right chemistry and the right age to be the mystery volcano. Westgate sent out word through the scientific grapevine that he needed ash samples of any really large volcanic explosions from the last ice age, which could be candidates for the eruption 74,000 years ago. First his lab group looked at the Laki volcano in Iceland, near the Greenland ice sheet. When it erupted in 1783, it spewed out more than 520 square kilometers (200 mi^2) of lava and ash, enough to cover Manhattan up to half the height of the Empire State Building. That eruption plunged the Northern Hemisphere into one of the coldest winters ever recorded. But Laki was the wrong age and the wrong chemistry for the mystery volcano. (Laki did have devastating impacts on people living close to the eruption: about 9,000 died in Iceland, about 25 percent of its population at the time, most of them succumbing to the ensuing famine.)

Many more samples were sent to Westgate, with the most likely candidates coming from Southeast Asia, a region with 70-plus major volcanoes. Among them is Mount Pinatubo in the Philippines, which has erupted many times in the past million years. In 1991, it exploded in spectacular fashion, blasting 10 to 15 billion tonnes (11 to 16.5 billion short tons) of glass and ash into the stratosphere, which moved around the entire earth in a matter of weeks. Despite warnings from geologists and the almost complete evacuation of the zone around the volcano, this eruption cost hundreds of Filipinos their lives and displaced thousands more into refugee camps.

But the 1991 Pinatubo ash, too, had the wrong chemistry to match the mystery samples. So did the ash from most of the rest of the known volcanoes of Southeast Asia, including Krakatau, which last erupted in 1883, and Tambora, which in 1815 ejected many times the amount of material that Krakatau later did and caused the "Year without a Summer" in 1816 (see chapter 3). Where was the missing megavolcano?

Mystery Solved

Finally, in the spring of 1994, Westgate got a package of ash in the mail from a new source. It came from a volcano that Craig Chesner of Eastern Illinois University had been studying for a long time. The volcano in question was a huge oval structure in northern Sumatra. Hundreds of meters high, 100 kilometers (62 mi) long, and 30 kilometers (20 mi) wide, it is now filled with the waters of Lake Toba, which plunges hundreds of meters deep (fig. 1.3). A ring of elevated terrain around the lake is so large that it can be seen from space, even without the high-resolution satellites of the kind that we use to gather images for Google Earth. The steep slope of the lakeshore is typical of the steep inner rim of a volcano, and indeed, this is the caldera created by a monstrous eruption larger than any in recorded history. Mount Saint Helens, by comparison, erupted catastrophically in 1980 and emitted a huge column of ash—but only 1 cubic kilometer (0.2 mi^3) of magma (see chapter 9). Toba's eruption produced many thousands of times as much ash as Mount Saint Helens and at least 10 times as much ash as any similar volcano in the region, such as Tambora or Krakatau, has emitted.

When Westgate analyzed Chesner's samples of Toba volcanic debris, the mystery was solved. The geochemistry and other characteristics of this ash clearly matched all the mystery samples from across the earth that had dates of around 74,000 years ago. The largest volcanic eruption in the past 28 million years finally had been found.

This amazing and important breakthrough was almost completely a product of serendipity. While combing through ice cores and analyzing the chemistry of ancient gas bubbles, Greg Zielinski had accidentally discovered evidence of a huge source of sulfuric acid dated 74,000 years ago. Independently, Mike Rampino had found evidence of catastrophic cooling at the same time, recorded by plankton fossils in deep-sea cores. These clues pointed to an unbelievably huge volcanic eruption, but no such eruption was yet known. Then John Westgate received samples from around the world of a mystery ash, all with the same date: 74,000 years ago. Finally, his call for help produced ash

Figure 1.3. Toba. *a.* Satellite image of Toba's oval caldera, now filled by the waters of Lake Toba, with the resurgent dome visible in the middle. *b.* Ground-level view of the lake that fills the Toba caldera. (*a:* Courtesy NASA; *b:* photograph by Martin Jones; courtesy Wikimedia Commons)

samples from Lake Toba that proved to be a match, and the first part of the puzzle was solved.

Before we look at the effects of the Toba eruption on the planet, however, we need to explore volcanoes in greater detail, including how they operate. That is the subject of the next chapter.

2 Vulcan's Fury

He also had one volcano that was extinct. But, as he said, "One never knows!" So he cleaned out the extinct volcano, too. If they are well cleaned out, volcanoes burn slowly and steadily, without any eruptions. Volcanic eruptions are like fires in a chimney.

On our earth we are obviously much too small to clean out our volcanoes. That is why they bring no end of trouble upon us.

Antoine de Saint-Exupéry, *The Little Prince*, 1944

The Forge of Hephaistos

To people of the ancient world, volcanoes were truly terrifying phenomena. The Greeks were awed by their tremendous power and imagined that they were the blazing forges where the god of fire, Hephaistos (known as Vulcan to the Romans), hammered out thunderbolts for Zeus. When eruptions occurred, it was said that Hephaistos was angry because his wife, Aphrodite (Venus), had cheated on him. In other Greek myths, volcanoes were said to be entrances to the underworld, ruled by the god Hades (Pluto).

In other parts of the world, eruptions were also seen as the work of the gods. The myths of Aztec culture tied the active volcanoes Popocatepetl and Iztaccihuatl, 70 kilometers (43 mi) southeast of Mexico City, to the legendary figures of those names. In one version of their myth, Popocatepetl was a great warrior who was sent off to battle with the promise that the chief's daughter, Iztaccihuatl, would be his in marriage if he returned. Later, Iztaccihuatl, believing that Popocatepetl had died, killed herself, and when the warrior returned, he killed himself, too (the scene is a bit like the ending of *Romeo and Juliet*, but in reverse order of death). According to Aztec legend, both lovers were laid out on high tables, which the gods turned into volcanoes as a sort of funeral pyre. Popocatepetl is the most active volcano in Mexico, having erupted at least 15 times since the Spanish arrived in 1519, most recently in March 2016.

In southern Italy, volcanoes are also a major part of the landscape and history. The word *volcano* itself comes from a small volcanic island north of Sicily that the Romans named Vulcano. Some volcanoes in this region, such as Mount Etna and the island of Stromboli, have erupted many times, but the most active and destructive is Mount Vesuvius, on the edge of the Bay of Naples. According to the Hellenistic Greek historian Diodorus Siculus, writing in 30 BCE in Sicily, it was part of the legend of Hercules (Herakles to the Greeks). While performing his twelve labors, the hero passed through the Phlegraean Plain (from the Greek for "to burn"), named for "a hill which anciently vomited out fire, . . . now called Vesuvius." Giant bandits, known as "sons of the earth," roamed the area, but Hercules quickly took care of them and went on to his next labor. Some scholars believe that the name Vesuvius may come from the Greek for "son of Zeus," since Hercules was one of Zeus's many children. The Romans, for their part, named the port city of Herculaneum in his honor.

Legends aside, in 79 CE the Romans were not concerned about Mount Vesuvius. Its most recent eruption had occurred in 217 BCE, almost 300 years earlier, and the event was lost in the distant mists of Roman history. The rich volcanic soil on its slopes was perfect for vineyards then, just as it is now, so Romans settled high up on the mountain, and many cities clustered on the Campanian Plain to the southeast, just as they do today. More than 20,000 people lived in Pompeii alone. Great earthquakes in 62 CE had damaged most of the area's cities, including Pompeii, Herculaneum, and Neapolis (now Naples), but the Romans rebuilt and ignored most of the ensuing temblors.

In August 79, the number of earthquakes increased. Even more ominously, some local springs and wells dried up, suggesting that the water table was dropping. August 23 was the feast of Vulcanalia, when the Romans honored Vulcan. The next afternoon, the god celebrated his festival in his own way. With a huge, deafening explosion of ash and gases, a mushroom cloud of ash rose above Vesuvius. The sky grew black, and ash and pumice rained down for about 18 hours. Although some people in the area fled as soon as they could, many were unable to leave because there were not enough boats in the harbor to evacuate them all and blankets of volcanic ash blocked the roads. When this first phase ended, the streets of Pompeii were buried under 2.8 meters (9.2 ft) of ash, and it was impossible to move anywhere or even breathe.

After almost a day of these hellish conditions, Vesuvius followed up with the main course. Huge eruptions of hot volcanic ash known as pyroclastic flows roared down the sides of the mountain. These clouds, superheated to about 850°C (1,600°F), incinerated nearly everything they touched and moved so fast (160 kph, or 100 mph) that nothing could escape them. When they subsided, a thick blanket of volcanic debris had almost completely buried Pompeii, Herculaneum, and the rest of the area. More than 20,000 people died in the eruption, with only a few thousand survivors managing to take boats away from the disaster.

There are very few firsthand accounts of deadly eruptions such as this one, and indeed most of the survivors of Vesuvius left no record. Fortunately, we do have one excellent account of this spectacular event. Only 17 at the time, Pliny the Younger—full name Gaius Plinius Caecilius Secundus—escaped with his family in a boat to the town of Misenum, about 35 kilometers (22 mi) across the Bay of Naples from the volcano. In a famous letter to a friend, the Roman historian Cornelius Tacitus, written 25 years after the eruption, he describes the death of his 56-year-old uncle Gaius Plinius Secundus (known to us as Pliny the Elder), one of Rome's leading admirals, scholars, and naturalists, who had decided to take a boat closer to the mountain to rescue his friends. I fondly remember translating this account in my high school Latin class, and its words remain vivid almost 2,000 years after they were written.

My dear Tacitus,

You ask me to write you something about the death of my uncle so that the account you transmit to posterity is as reliable as possible. I am grateful to you, for I see that his death will be remembered forever if you treat it [in your *Histories*]. He perished in a devastation of the loveliest of lands, in a memorable disaster shared by peoples and cities, but this will be a kind of eternal life for him. Although he wrote a great number of enduring works himself, the imperishable nature of your writings will add a great deal to his survival. Happy are they, in my opinion, to whom it is given either to do something worth writing about, or to write something worth reading; most happy, of course, those who do both. With his own books and yours, my uncle will be counted among the latter. It is therefore with great pleasure that I take up, or rather take upon myself, the task you have set me.

He was at Misenum in his capacity as commander of the fleet on the 24th of August [79], when between 2 and 3 in the afternoon my mother drew his attention to a cloud of unusual size and appearance. He had had a sunbath, then a cold bath, and was reclining after dinner with his books. He called for his shoes and climbed up to where he could get the best view of the phenomenon. The cloud was rising from a mountain—at such a distance we couldn't tell which, but afterward learned that it was Vesuvius. I can best describe its shape by likening it to a pine tree [today we would compare it to a mushroom cloud]. It rose into the sky on a very long "trunk" from which spread some "branches." I imagine it had been raised by a sudden blast, which then weakened, leaving the cloud unsupported so that its own weight caused it to spread sideways. Some of the cloud was white, in other parts there were dark patches of dirt and ash. The sight of it made the scientist in my uncle determined to see it from closer at hand. [This kind of explosive mushroom cloud of ash and pumice is now called a Plinian eruption in his honor.]

He ordered a boat made ready. He offered me the opportunity of going along, but I preferred to study—he himself happened to have set me a writing exercise. As he was leaving the house he was brought a letter from Tascius [Pomponianus Publius]'s wife, Rectina, who was terrified by the looming danger. Her villa lay at the foot of Vesuvius, and there was no way out except by boat. She begged him to get her away. He changed his plans. The expedition that started out as a quest for knowledge now called for courage. He launched the quadriremes [galleys with four banks of oars apiece] and embarked himself, a source of aid for more people than just Rectina, for that delightful shore was a populous one. He hurried to a place from which others were fleeing, and held his course directly into danger. Was he afraid? It seems not, as he kept up a continuous observation of the various movements and shapes of that evil cloud, dictating what he saw.

Ash was falling onto the ships now, darker and denser the closer they went. Now it was bits of pumice, and rocks that were blackened and burned and shattered by the fire. Now the sea is shoal; debris from the mountain blocks the shore. He paused for a moment wondering whether to turn back as the helmsman urged him. "Fortune helps the brave," he said. "Head for Pomponianus."

At Stabiae, on the other side of the bay formed by the gradually curving shore, Pomponianus had loaded up his ships even before the danger arrived, though it was visible and indeed extremely close, once it intensified. He planned to put out as soon as the contrary wind let up. That very wind carried my uncle right in, and he embraced the frightened man and gave him comfort and courage. In order to lessen the other's fear by showing his own unconcern, he asked to be taken to the baths. He bathed and dined, carefree or at least appearing so (which is equally impressive). Meanwhile, broad sheets of flame were lighting up many parts of Vesuvius; their light and brightness were the more vivid for the darkness of the night. To alleviate people's fears my uncle claimed that the flames came from the deserted homes of farmers who had left in a panic with the hearth fires still alight. Then he rested, and gave every indication of actually sleeping; people who passed by his door heard his snores, which were rather resonant since he was a heavy man. The ground outside his room rose so high with the mixture of ash and stones that if he had spent any more time there escape would have been impossible. He got up and came out, restoring himself to Pomponianus and the others who had been unable to sleep. They discussed what to do, whether to remain under cover or to try the open air. The buildings were being rocked by a series of strong tremors, and appeared to have come loose from their foundations and to be sliding this way and that. Outside, however, there was danger from the rocks that were coming down, light and fire-consumed as these bits of pumice were. Weighing the relative dangers, they chose the outdoors; in my uncle's case it was a rational decision, others just chose the alternative that frightened them the least.

They tied pillows on top of their heads as protection against the shower of rock. It was daylight now elsewhere in the world, but there the darkness was darker and thicker than any night. But they had torches and other lights. They decided to go down to the shore, to see from close up if anything was possible by sea. But it remained as rough and uncooperative as before. Resting in the shade of a sail he drank once or twice from the cold water he had asked for. Then came a smell of sulfur, announcing the flames, and the flames themselves, sending others into flight but reviving him. Supported by two small slaves he stood up, and immediately collapsed. As I understand it, his breathing was obstructed by the dust-laden air, and his innards, which were never strong and often blocked or upset, simply shut down. When daylight came again two days after he died, his body was found untouched, unharmed, in the clothing that he had had on. He looked more asleep than dead.

In a second letter to Tacitus a few days later, Pliny wrote:

By now [the morning after the eruption began] it was dawn, but the light was still dim and faint. The buildings round us [Pliny and his mother] were already tottering, and the open space we were in was too small for us not to be in real and imminent danger if the house collapsed. This finally decided us to leave the town. We were followed by a panic-stricken mob of people wanting to act on someone else's decision in preference to their own (a point in which fear looks like prudence), who hurried us on our way by pressing hard behind in a dense crowd. Once beyond the buildings we stopped, and there we had some extraordinary experiences which thoroughly alarmed us. The carriages we had ordered to be brought out began to run in different directions though the ground was quite level, and would not remain stationary even when wedged with stones. We also saw the sea sucked away and apparently forced back by the earthquake: at any rate it receded from the shore so that quantities of sea creatures were left stranded on dry sand. On the landward side a fearful black cloud was rent by forked and quivering bursts of flame, and parted to reveal great tongues of fire, like flashes of lightning magnified in size.

At this point, my uncle's friend from Spain spoke up still more urgently: "If your brother, if your uncle is still alive, he will want you both to be saved; if he is dead, he would want you to survive him—why put off your escape?" We replied that we would not think of considering our own safety as long as we were uncertain of his. Without waiting any longer, our friend rushed off and hurried out of danger as fast as he could.

Soon afterward the cloud sank down to earth and covered the sea; it had already blotted out Capri and hidden the promontory of Misenum from sight. Then my mother implored, entreated, and commanded me to escape the best I could—a young man might escape, whereas she was old and slow and could die in peace as long as she had not been the cause of my death too. I refused to save myself without her, and grasping her hand forced her to quicken her pace. She gave in reluctantly, blaming herself for delaying me. Ashes were already falling, not as yet very thickly. I looked round: a dense black cloud was coming up behind us, spreading over the earth like a flood. "Let us leave the road while we can still see," I said, "or we shall be knocked down and trampled underfoot in the dark by the crowd behind." We had scarcely sat down to rest when darkness fell, not the dark of a moonless or cloudy night, but as if the lamp had been put out in a closed room. You could hear the shrieks of women, the wailing of infants, and the shouting of men; some were calling their parents, others their children or their wives, trying to recognize them by their voices. People bewailed their own fate or that of their relatives, and there were some who prayed for death in their terror of dying. Many besought the aid of the gods, but still more imagined there were no gods left, and that the universe was plunged into eternal darkness for evermore. There were people, too, who added to the real perils by inventing fictitious dangers: some reported that part of Misenum had collapsed or another part was on fire, and though their tales were false they found others to believe them. A gleam of light returned, but we took this to be a warning of the approaching flames rather than daylight. However, the flames remained some distance off; then darkness came on once more and ashes began to fall again, this time in heavy showers. We rose from time to time and shook them off, otherwise we should have been buried and crushed beneath their weight. I could boast that not a groan or cry of fear escaped me in these perils, had I not derived some poor consolation in my mortal lot from the belief that the whole world was dying with me and I with it.

At last the darkness thinned and dispersed into smoke or cloud; then there was genuine daylight, and the sun actually shone out, but yellowish as it is during an eclipse. We were terrified to see everything changed, buried deep in ashes like snowdrifts. We returned to Misenum, where we attended to our physical needs as best we could, and then spent an anxious night alternating between hope and fear. Fear predominated, for the earthquakes went on, and several hysterical individuals made their own and other people's calamities seem ludicrous in comparison with their frightful predictions. But even then, in spite of the dangers we had been through, and were still expecting, my mother and I had still no intention of leaving until we had news of my uncle.

By the time the eruption ended, more than 20 meters (66 ft) of ash covered Pompeii, so it was almost completely buried. Later another Roman town was built right on top of it, and the old city was lost to memory. But eventually, in 1748, well diggers found some of the ruins, and over the next 150 years most of the town was excavated (and parts of it still are being worked on today). It is a time capsule of life in the early Roman Empire, with brilliant frescoes, mosaic floors, and even graffiti on the walls. Even more remarkable were the hollow cavities in the ash that workers discovered as they dug inside buildings. Archaeologists filled these cavities with plaster, chipped away the surrounding ash, and discovered that they had casts of bodies: those of people and their dogs who had died in the eruption, curled up in the fetal position as they asphyxiated in the searing ash and then were incinerated. Even as they vaporized, the fluid in their bodies chilled and consolidated the ash around them.

Like Pompeii, Herculaneum was entombed, but in an even thicker blanket of dense, hard, welded volcanic ash called tuff instead of the softer debris that had buried Pompeii. Even though it was rediscovered in 1738, the tuff there is so hard that only about 20 percent of the town has been exposed nearly three centuries later. In its own day, Herculaneum was a small but affluent coastal resort of about 5,000 people; the wealth of its residents is indicated by archaeologists' discovery there of clothing and jewelry that are far more ornate than those found in Pompeii. Excavators at Herculaneum found not only hollow cavities in the shape of human corpses but also 300 skeletons in death poses. Most were near

Figure 2.1. Vesuvius erupting in 1944 as US B-25 Mitchell bombers fly near the ash plume. (Courtesy World War II Database)

the waterfront, where people had been trapped, unable to find a boat to take them to safety.

The eruptions of Vesuvius did not end with the catastrophe of 79 CE. The Roman historian Cassius Dio recorded another one in 203, and the eruption of 472 spewed ash that fell as far away as Constantinople (modern Istanbul, Turkey). In 1906, a huge eruption produced a record number of lava flows and killed more than 100 people. It so devastated the city of Naples that the Italian government's diversion of funds to rebuild the town forced Rome to back out of hosting the 1908 Summer Olympics. The most recent major eruption, in March 1944, destroyed several villages. It also damaged 88 B-25 Mitchell bombers of the US Air Force's 340th Bombardment Group, which was stationed near Pompeii as part of the Allied effort to drive the Germans out of Italy during World War II (fig. 2.1). Since then, Vesuvius has gone dormant again, although steam still rises from its crater. Today more than 3 million people live around its base and a million more on its slopes. Like humans everywhere, they ignore the hazard of geological catastrophes that happen only every few decades or centuries, taking advantage of good conditions before the next disaster arrives.

Incineration

The memories of Vesuvius would have been lost to history were it not for the account of Pliny the Younger, which survives as perhaps one of the first scientific descriptions of a volcano in action. Both Plinies treated the eruption as a natural phenomenon to be studied and recorded, not as some act of an angry deity to whom they should pray for mercy. Yet even volcanologists reading Pliny the Younger's account just a little over a century ago could not fully visualize the horrors of the pyroclastic flows or the blinding, suffocating, scalding rain of ash, because few scientists had ever experienced such things and lived to tell the tale.

Our understanding of such catastrophic conditions developed thanks largely to an incredible eruption in the New World. The islands of the Caribbean include many active volcanoes, such as the one in Montserrat's Soufrière Hills, which has erupted as recently as 2012. But the deadliest of all the recorded Caribbean eruptions occurred on the island of Martinique in 1902.

Back then, Martinique was a densely populated French colony, with a port city, Saint-Pierre, that had a population of more than 25,000 people. Saint-Pierre dated all the way back to 1635 and traded in many tropical goods (especially sugarcane), supporting a very cultured lifestyle for its inhabitants. Its nickname was "the Paris of the Caribbean." Yet it was no stranger to natural disasters. The Great Hurricane of 1780 had pounded Saint-Pierre with a wall of water (known as a storm surge) that was 8 meters (25 ft) deep. This inundation killed 9,000 people and destroyed most of the city's buildings.

Saint-Pierre was built along the island's northern shore, below the slopes of its main volcano, Mount Pelée, which means "peeled mountain" or "bald mountain" in French. Mount Pelée had been dormant for centuries, so most people did not regard it as a threat. However, the local Carib people were familiar with its past and called it Fire Mountain.

By early 1902, it was clear that something was happening to the mountain. On April 23, sightseers noticed fumaroles emerging from its top, but no one was alarmed, because this had happened often in the past. On the same day, Mount Pelée spewed a light rain of cinders onto its southern and western flanks, accompanied by several strong earthquakes. Two days later, a large cloud of ash and rocks blew out of the top but didn't cause much damage. The next day, another dusting of volcanic ash covered Saint-Pierre. Still the residents saw no cause for worry, because the mountain had done these things before without erupting. They were also overconfident that even if lava flows should erupt, they would follow two large river valleys to the sea, bypassing the city. In 1902, no one knew that volcanoes could erupt explosively and produce huge clouds of ash and pyroclastic flows that can easily travel over such valleys.

On April 27, sightseers who climbed to the top of Mount Pelée found that its main crater, Étang Sec, had filled with water, forming a lake more than 180 meters (600 ft) across. On one side was a cone of volcanic debris more than 15 meters (50 ft) high, which was pouring boiling water into the lake; rumblings were heard underground, and sulfurous fumes filled Saint-Pierre. By April 30, two of the local rivers were flooded, their currents grown strong enough to carry boulders and trees away from the mountain. On May 2, Mount Pelée produced a huge pillar of black smoke, along with explosions and frequent earthquakes. Eruptions at five- to six-hour intervals covered the northern half of the island with ash. The local newspaper, *Les Colonies*, had to postpone a picnic on the mountain scheduled for May 4, and pastured animals began to die as ash covered the ground and tainted their water supplies.

Still the mayor of Saint-Pierre assured everyone that there was no need to worry. No evacuation was necessary, he said; the mountain had threatened them before but had never erupted fully, and besides, the two valleys would divert any lava flows away from the city. His concern was for the upcoming election, in which he was running—he did not want the voters to flee before it was over. As it happened, things did seem less dire on May 3, when the wind shifted and blew the ash cloud northward, making life better for a while. But the next day the ashfall got worse, cutting off communications. Coastal boats were afraid to approach Saint-Pierre through the ash, and people filled every available departing boat to capacity. Still, only a tiny portion of the 28,000 people in the area were able to leave.

On May 5, the mountain seemed quieter. But in the early afternoon, the sea receded about 100 meters (330 ft) from the shoreline and then rushed back, flooding parts of Saint-Pierre in a tsunami apparently caused by one of the frequent earthquakes beneath Mount Pelée. Up on the mountain itself, the crater rim collapsed into the new lake, and a lahar (a Javanese term for a boiling mudflow) poured over the volcano's lip and into the Blanche River. About 150 people were buried under 90 meters (300 ft) of mud as a local sugar works flooded. Meanwhile, people from towns nearer the mountain fled to the relative safety of Saint-Pierre. But the electrical grid failed because of atmospheric disturbances caused by ash, plunging the city into eerie darkness as the mountain continued to rumble in the distance.

That same day, a host of bizarre events happened that seemed portents of doom. Livestock in the fields were distressed and agitated and tried to jump their fences. At a sugar mill north of the city, the ground swarmed with yellowish speckled ants and 30-centimeter- (1-ft-) long black centipedes, driven from the slopes of Mount Pelée by constant tremors and ash. They bit animals and people, although none died from these attacks. Snakes invaded the streets of Saint-Pierre, including 2-meter (6-ft) poisonous pit vipers known as fer-de-lance. The mayor of Saint-Pierre sent soldiers out to shoot them, and at least 100 snakes were killed, but not before they had bitten and killed more than 200 animals (mostly pigs and chickens) and 50 people.

Saint-Pierre residents woke at 4 a.m. on May 7 to bigger rumblings, more explosions, and an orange-red glow on the mountain to the north. By daybreak, the clouds of ash were generating their own weather, making lightning that struck the mountain over and over again. The mayor and the newspapers assured the people that they were safe. The news that

La Soufrière volcano, on the nearby island of Saint Vincent, was also erupting made them think that it was venting the pressure from Mount Pelée. Still, more people tried to leave; Captain Marina Leboffe sailed off that day with only half of the sugar cargo loaded onto his ship *Orsolina*, even though his shipping company protested and port authorities refused to grant him clearance. Others trying to leave were threatened with arrest. In addition to the mayor, the French colonial governor Louis Mouttet and his wife made a big show of staying in Saint-Pierre to reassure the population. Their bravado seemed justified, because the mountain quieted down again that night.

May 8 was Ascension Day, when Catholics celebrate the ascension of Jesus into heaven (the 40th day after Easter). In Saint-Pierre in 1902, it dawned with something else rising to heaven. At 7:52 a.m., the last telegraph transmission from observers near the mountain arrived, reading, "Allez" (Go). Then the line went dead. The next second, Mount Pelée exploded, generating a huge plume of ash that shot up into the stratosphere. One surviving observer, who was on a boat at the time of the eruption, said the mountain blew apart without warning, while another compared it to the explosion of a giant oil refinery. A second, larger eruption formed a mushroom cloud that darkened the sky for 80 kilometers (50 mi) in all directions. But even deadlier was the surge of ash that flowed horizontally from the rim of the volcano, hugging the ground as it went. It traveled up to 670 kilometers (420 mi) per hour, arriving in Saint-Pierre just seconds after it had started. The two river valleys proved to be no protection for the city, because this was an entirely different type of eruption: one without lava flows. This was the first time that modern observers had seen pyroclastic flows in action; they were soon dubbed *nuées ardentes* ("glowing clouds" in French). When these glowing clouds hit Saint-Pierre, the citizens were suffocated and then burned to death by volcanic gases reaching 1,075°C (1,967°F). All the flammable buildings and other objects in town and all the ships in port also burst into flames and vanished in minutes (fig. 2.2).

The steamship *Roraima* had reached the port of Saint-Pierre just minutes before the eruption and was engulfed in the pyroclastic flow. It

Figure 2.2. *a.* Saint-Pierre, Martinique, before the Mount Pelée eruption of 1902, with the volcano in the background. *b.* Saint-Pierre afterward, burned nearly to the ground, with only masonry walls standing. (Photographs by Angelo Heilprin, 1902; courtesy Wikimedia Commons)

burned and then sank, killing the entire crew of 28 and all passengers except a little girl and her nurse. The wreck still lies on the seafloor just outside the modern city of Saint-Pierre. Almost all those who survived the initial blast—and they were few in number—were so horribly burned that they died days later. Before she died, one of the survivors spoke of a sudden blast of extreme heat but remembered nothing else about the eruption.

Officially, only two of the more than 30,000 people who were in Saint-Pierre at the time are known to have still been alive more than a year later. One was a felon, Louis-Auguste Cyparis, who was saved because he was imprisoned in a deep, windowless underground cell that the superheated ash could not reach. When he was rescued a few days later, he was barely alive, but he recovered from his injuries and spent the rest of his life as a celebrity in the sideshow of the Barnum & Bailey circus, displaying the burn scars that covered his body.

The other survivor, Léon Compère-Léandre, tried to flee the eruption but had barely left his home, on the edge of town, before the blast over-took him. On June 26, 1902, the Parisian newspaper *Le Temps* published his description of his experience:

> I felt a terrible wind blowing, the earth began to tremble, and the sky suddenly became dark. I turned to go into the house, with great diffi-culty climbed the three or four steps that separated me from my room, and felt my arms and legs burning, also my body. I dropped upon a table. At this moment four others sought refuge in my room, crying and writhing with pain, although their garments showed no sign of having been touched by flame. At the end of 10 minutes one of these, the young Delavaud girl, aged about 10 years, fell dead; the others left. I got up and went to another room, where I found the father Delavaud, still clothed and lying on the bed, dead. He was purple and inflated, but the clothing was intact. Crazed and almost overcome, I threw myself on a bed, inert and awaiting death. My senses returned to me in perhaps an hour, when I beheld the roof burning. With suf-ficient strength left, my legs bleeding and covered with burns, I ran to Fonds-Saint-Denis, six kilometers [4 mi] from Saint-Pierre.

For the next few hours, rushes of hot wind covered Saint-Pierre, along with a downpour of muddy ash. Wooden ships approaching the ghostly seaport were turned back by the heat of the ash, not wanting to risk catching fire. At 12:30 p.m. on the day of the eruption, the acting governor of Martinique sent the iron warship *Suchet* to investigate, but it, too, was turned back by the heat until late afternoon. When relief efforts at last arrived in the following days, they found the entire city and its surroundings covered with volcanic ash. Nearly every building was burned down to the masonry, every plant was blackened, and thousands of scorched corpses were rotting in the sun. So many buildings and landmarks had disappeared that rescuers could not recognize where they were.

Initially, the fires, the intense heat, and the stench of death prevented rescuers from going much past the shoreline. In the next few days, as they slowly made their way into the city, they found that a zone of about 20 square kilometers (8 mi²) had been totally destroyed. Surrounding it, another zone had lost many people, but fewer buildings had burned. Many of the dead were buried meters deep under volcanic ash, and their bodies were not retrieved, since most were unidentifiable. Among the numerous corpses, some were found in relaxed attitudes, with calm features suggesting that the blow had hit them with no warning and killed them instantly. Others of the dead, though, were contorted in anguish. Those found outdoors had had their clothing burned completely away. The city continued to smolder for days, and as the fires cooled, the rescue crews disposed of thousands of remaining corpses by incinerating them as well, since there was no time for burial. The smell of the rotting bodies was, rescuers reported, unbelievable.

Many nations sent ships to aid the relief effort, but Mount Pelée was not yet done. On May 20, a second eruption, just as powerful as the May 8 event, obliterated the remains of Saint-Pierre, killing 2,000 rescuers, engineers, and sailors. Still another powerful eruption, on August 30, 1902, sent pyroclastic flows even farther east than those of the initial eruptions. They hit a number of towns that had escaped the first two events, killing another 1,100 people and causing a tsunami that damaged the coastal town of Le Carbet.

The cataclysmic events on Mount Pelée marked the beginning of modern volcanology and prompted the realization that not all eruptions are simple lava flows like those that built Hawaii (see the next section). The volcanologists Angelo Heilprin and Antoine Lacroix published extensive reports on this case, adding the term *nuée ardente* to their field. And just as Vesuvius-style eruptions are now called Plinian, those dominated by pyroclastic flows are today known to volcanologists as Pelean.

Why Do Volcanoes Behave So Differently from One Another?

The examples of Mount Vesuvius and Mount Pelée show that there are important differences among volcanic events. Whereas Hawaii's volcanoes erupt relatively quietly and have killed relatively few people, the victims of Martinique learned the hard way that river valleys might divert lava flows but do not protect against fast-moving, ground-hugging, superheated clouds of volcanic ash and gases that can travel across obstacles for many kilometers.

The difference lies in the chemistry of the magma, the molten rock that erupts to form lava. Lavas like those in Hawaii or Iceland form when hot plumes of magma rise up from the earth's mantle or when the crust of the midocean ridges rips open deep on the ocean floor. These mantle-derived magmas are rich in elements such as magnesium and iron, and the resulting black rocks that form after the lava has cooled are called basalt (pronounced "bah-SALT" by American geologists and "BASS-alt" by British geologists). Magnesium- and iron-rich magmas melt at very high temperatures (typically 1,600°C, or about 2,900°F), so they are very hot but also very fluid. They erupt as lava that flows along the landscape with the viscosity of water. Such magmas seldom explode but can throw blobs of lava in the air while erupting. In most cases, people can get out of the way of these lava flows, which don't move very fast. People on the slopes of Kilauea on the Big Island of Hawaii, for example, must often cope with the destruction caused by such flows but are rarely in any personal danger from them (although their immovable property is sometimes burned up or covered by black lava).

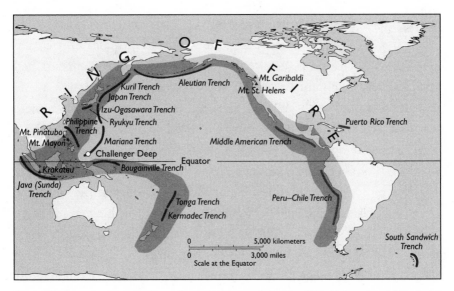

Figure 2.3. Zones of major earthquakes and explosive volcanoes in the Ring of Fire. (Redrawn from US Geological Survey data)

At the other extreme are magmas that are richer in silicon, aluminum, sodium, and potassium. These elements compose minerals that melt at lower temperatures than basaltic minerals (sometimes as low as 600°C, or 1,000°F). Depending on their exact chemistry, these magmas are known as andesites (after the Andes Mountains), dacites, or rhyolites. They tend to be very viscous and flow like molasses or peanut butter. Especially in the case of rhyolite, they are too thick and sticky to generate any lava flows at all. Instead, these magmas plug their volcanoes until so much pressure builds up underneath them that eventually the top of the mountain simply explodes. These explosions produce the wide range of shattered volcanic rock known as pyroclastics (from the Greek for "fire" and "fragments"), from tiny glass shards of volcanic ash to the pebble-size pieces of pumice known as lapilli. The largest eruptive materials, cobble- and boulder-size blobs of magma that cool as they fly through the air, are appropriately known as volcanic bombs.

Andesite, dacite, and rhyolite magmas began as basaltic magma but were transformed into entirely new chemical compositions in subduction zones: places where one tectonic plate plunges under another. A downgoing plate slides beneath an overlying plate until it reaches a depth where it is so hot that the material in the plate (and the base of the crustal plate above it) begins to melt, forming magma. Such melts are enriched with sodium, aluminum, potassium, and silica because the minerals made of those elements melt at lower temperatures than do basaltic minerals. As the rocks in the downgoing and overlying plates heat up, these low-melting-temperature minerals melt first, producing these distinctive magmas. Blobs of the magmas then rise to the surface, sometimes absorbing even more sodium, aluminum, potassium, and silica as they melt the overlying crustal rocks, made of granites and metamorphic rocks that are rich in those elements. Finally, they reach the surface, where they eventually form a chain of volcanoes on the face of the earth as the tectonic plates continue to move. Because of their magma chemistry, subduction-zone volcanoes are almost always explosive and generate pyroclastic eruptions, with few or no lava flows.

Most of the earth's subduction zones are found around the rim of the Pacific Ocean, arranged roughly in a circle nicknamed the Ring of Fire. This is the location of not only most of the world's deadliest volcanoes but also most of its biggest earthquakes (fig. 2.3). The latter occur when a downgoing plate slips suddenly as it grinds past an overlying slab. When you look at a map of the Ring of Fire, you will see long chains or arcs of volcanoes. Some appear as volcanic islands in the middle of the ocean, like those found in Japan, the Aleutians, the Philippines, and Indonesia. Others sit on the edges of continents, where an oceanic plate is plunging beneath a thicker continental plate. These land-based volcanoes include the Andes, the Central American volcanic arc, and the Cascade Mountains of Northern California, Oregon, Washington, and British Columbia.

There is far more to volcanology than this brief introduction can cover, but this is not a textbook. In the next chapter, we will look at the Indonesian volcanic arc in detail and how Toba fits into the picture.

3

Land of the Killer Volcanoes

Mount Kilauea spilled glowing lava like cords of orange neon-lighting from seemingly nowhere. In the blackness that engulfed the night, electric heat lit flowing streams that fell into the sea, disappearing in a cloud of steam with a sizzling splash.

Victoria Kahler, *Capturing the Sunset*, 2011

The Indonesian Arc

There are volcanoes all around the Ring of Fire and in many other places on earth, but some of the most impressive and violent eruptions in history have occurred in the chains of islands that are known as the Malay Archipelago, which includes Indonesia and parts of Malaysia. Moreover, this volcanic chain harbors almost continuous eruptions. Indonesia alone has more volcanoes than any other country in the world, including about 150 major ones—127 of them still active—and hundreds more that are extinct and weathered (fig. 3.1). In total, Indonesian volcanoes have erupted more than 1,171 times in recorded history. Consequently, constant volcanic activity affects nearly the whole country.

The size of volcanic eruptions is generally measured with the Volcanic Explosivity Index, or VEI, which is based on the amount of debris scattered during an explosion (fig. 3.2). Most Hawaiian volcanoes, which erupt lots of fluid lava in flows but seldom explode or blow debris high into the air, have a VEI rating of 0 to 1. The deadly 1985 eruption of Nevado del Ruiz in Colombia and the 1914–17 eruption of Mount

Figure 3.1. Major volcanoes of Indonesia. (Redrawn from US Geological Survey data)

Lassen in California each rated a 3, while Mount Pelée, in 1902, was a typical 4. Mount Vesuvius in 79 CE and Mount Saint Helens in 1980 had VEIs of 5. Like the Richter scale, used to measure earthquakes, the VEI is logarithmic, so a VEI 5 eruption (with more than 1 km³, or 0.2 mi³, of ejecta) produces 10 times as much material as a VEI 4 (more than 0.1 km³) and 100 times as much as a VEI 3 (more than 0.01 km³). The eruption of Krakatau in 1883 rated a 6 (with more than 10 km³ of ejecta), as did the explosion of Mount Pinatubo in the Philippines in 1991.

The two Indonesian volcanoes most active today are Kelut and Merapi, both on the island of Java. They have each killed thousands of people over the years. In the past thousand years, Kelut has erupted more than 30 times, with several of those eruptions rating a VEI 5. Merapi has erupted more than 80 times over the same period. Kelut's most recent eruptions were in 1919, 1951, and 1966 and on February 10, 1990, when it killed 35 people. Mount Merapi last erupted on November 3, 2010, killing 353. By the time this book is published, one or both of these volcanoes probably will have erupted again.

Unlike most of the rest of the volcanoes in the Ring of Fire, which are caused by some portion of the Pacific seafloor subducting beneath an adjacent continental plate, the volcanic arc in western Indonesia is a result of the Indian oceanic plate plunging beneath the island chain from the west. Nearly all of the world's tsunamis happen in the Pacific Ocean,

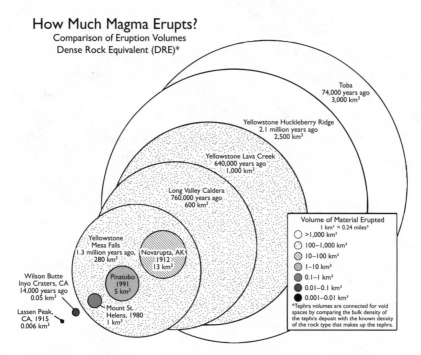

Figure 3.2. Eruptions of differing amounts of magma. Circle size represents the volume of ejected material. (Courtesy US Geological Survey)

since the Pacific has most of the world's subduction zones. Consequently, the countries of the Pacific Rim have long had tsunami warning systems in place, which send out alerts within seconds of any giant earthquake around the Pacific. However, because the subduction zone beneath Java and Sumatra is one of the few in the Ring of Fire that are not on the Pacific, tsunamis are comparatively rare in that region, and it had no system to warn about them on December 26, 2004, when a quake measuring 9.1 to 9.3 on the Richter scale occurred off the northwestern tip of Sumatra. This is why there were so many deaths during the ensuing Indian Ocean tsunami—more than a quarter million.

Indonesia has a huge population—more than 264 million in 2018—making it the world's fourth-largest country in population. As the entire island chain is built from volcanoes, Indonesians have little choice but to live near them; about 5 million people reside within the danger zone of one volcano or another. The rich volcanic soils and tropical climate allow people to grow a huge variety of crops, yielding products such as rubber, palm oil, coconut oil, cocoa, and coffee. Growing such crops requires many Indonesians to work close to active volcanoes as well (and has encouraged the deforestation of much of Indonesia's once-diverse rain forest).

The government of Indonesia has taken volcanoes seriously for a long time, as do Indonesians themselves. The Volcanological Survey of Indonesia was founded back in 1920 to monitor and help predict the likelihood of volcanic events, warn people about them, and issue orders for evacuation. These jobs are important. Before the 1982 eruption of Galunggung, which was a VEI 4, the agency ordered 75,000 people to evacuate; in the end, only 68 died. Signs of the forthcoming eruption of Makian in 1988 triggered the evacuation of all 15,000 people from its island, and no one died. In the past, before such evacuations, death tolls were huge—as I discuss below, more than 90,000 people died in the 1815 eruption of Tambora and over 36,000 in the 1883 eruption of Krakatau. No one can afford to dismiss such warnings today.

Krakatau Cracks

At the start of the 1880s, the western Indonesian archipelago was a quiet part of the Dutch East Indies colony, with little memory of giant volcanic eruptions, since there had been few in the previous 200 years. Planters and other colonists were trying to create a small version of Holland in the sweltering tropics. They ruled a huge native population who produced nearly all the world's supply of pepper and quinine, a third of its rubber, a quarter of its coconuts, and a fifth of its tea, cocoa, sugar, coffee, and oil, as well as other commodities needed around the earth. The colonists knew that there were dangerous volcanoes in Indonesia but had learned to cope with them, as they had with all the other local hazards, from tropical diseases to earthquakes.

Then, in 1882 and 1883, numerous earthquakes centered beneath the small island of Krakatau signaled the awakening of this quiet region. Krakatau (Indonesian spelling; the traditional English spelling is Krakatoa) was a long-dormant three-cone volcano in the center of the Sunda Strait between the islands of Sumatra and Java. Heavy ship traffic passed through the strait to and from Batavia (now Djakarta), on the north shore of Java, and between the Indian and Pacific Oceans. Krakatau had most recently erupted in 1680.

On May 20, 1883, huge plumes of steam rose from Perboewatan, the northernmost of Krakatau's three cones (fig. 3.3). Soon ash began exploding from the vent, rising to an altitude of 6 kilometers (4 mi), and the noise of the eruption could be heard as far away as Batavia, more than 160 kilometers (100 mi) distant. Volcanic activity died down for a few weeks thereafter, but it resumed on June 16 with more explosions. For five days, a thick ash cloud darkened the sky over Indonesia. Steady winds blew it away on June 24, allowing observers to see that there were two ash plumes, both coming from vents near the center of the volcano, south of Perboewatan. This phase of the eruption made the tides rise and fall erratically, so ships in ports around the Sunda Strait had to be chained to docks. Daily earthquakes rocked the islands. Even ships passing west of Krakatau reported large rafts of pumice floating in the Indian Ocean.

On August 11, the Dutch topographical engineer H. J. G. Ferzenaar landed on the island of Krakatau and conducted the only study of the early phase of activity before the final cataclysm. All vegetation had burned down, leaving just tree stumps, and there were now three major ash plumes, the new one coming from the central peak, Danan. Steam plumes rose from 11 other vents, and an ash layer 50 centimeters (20 in) deep covered the entire island. Ferzenaar reported on what he saw and recommended that no one else visit Krakatau, although life continued as normal in distant cities such as Batavia.

Meanwhile, the vents continued spewing ash, pumice, and steam. Then, on August 25, a big eruption vomited a cloud of black ash 27 kilometers (17 mi) high. More explosions followed at intervals of about 10 minutes. Hot ash and pumice up to 10 centimeters (3 in) in diameter landed on the decks of ships 20 kilometers (11 mi) distant. A small tsunami hit the shores of Java and Sumatra up to 40 kilometers (28 mi) away after one of Krakatau's volcanic cones collapsed into the sea. People in far-off Batavia could not sleep thanks to the eruption's deafening, pounding noise, which shattered the stillness of the night.

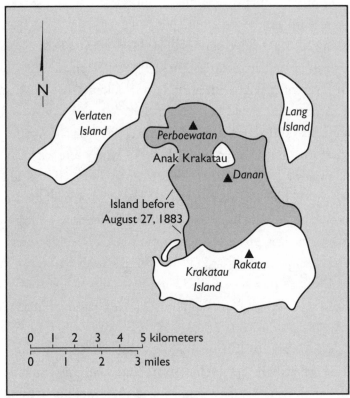

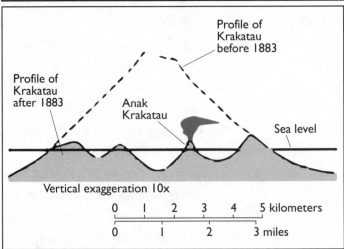

Figure 3.3. Original footprint of Krakatau and the remnant islands that formed after its 1883 eruption (top). In the lower diagram, Krakatau's original profile is contrasted with the tiny islands on the rim of the caldera, now flooded by the ocean. Anak Krakatau has grown in the old caldera since the eruption. (Redrawn from US Geological Survey data)

The climax came on the morning of August 27 with four large explosions between 5:30 and 10:41 a.m. They generated a huge pressure wave that spread from the volcano at 1,086 kilometers (675 mi) per hour. Its volume was measured at 310 decibels, loud enough to be heard in Perth, Australia, more than 3,500 kilometers (2,200 mi) away, and even on the island of Mauritius, north of Madagascar, 4,800 kilometers (3,000 mi) away, where people reported what sounded like nearby cannon fire. The pressure wave was so strong that it ruptured the eardrums of sailors in the Sunda Strait. Barometric gauges, which measure air pressure, jumped off their scales and shattered. This shock wave continued around the globe and was registered on barographs worldwide up to seven days after the initial explosions.

About 1,000 people died in the village of Ketimbang, more than 40 kilometers (25 mi) north of Krakatau, as burning ash rained down over the region. The wife of Controller Willem Beyerinck, the Dutch official responsible for the district, would later describe her experiences there as pyroclastic flows trapped and killed most of the residents:

Suddenly, it became pitch dark. The last thing I saw was the ash being pushed up through the cracks in the floorboards, like a fountain. I turned to my husband and heard him say in despair, "Where is the knife? . . . I will cut all our wrists and then we shall be released from our suffering sooner." The knife could not be found. I felt a heavy pressure, throwing me to the ground. Then it seemed as if all the air was being sucked away and I could not breathe. . . . I felt people rolling over me. . . . No sound came from my husband or children. . . . I remember thinking, I want to . . . go outside . . . but I could not straighten my back. . . . I tottered, doubled up, to the door. . . . I forced myself through the opening. . . . I tripped and fell. I realized the ash was hot and I tried to protect my face with my hands. The hot bite of the pumice pricked like needles. . . . Without thinking, I walked hopefully forward. Had I been in my right mind, I would have understood what a dangerous thing it was to . . . plunge into the hellish darkness. . . . I ran up against . . . branches and did not even think of avoiding them. I entangled myself more and more. . . . My hair got caught up. . . . I noticed for the first time that [my] skin was hanging off everywhere, thick and moist from the ash stuck to it. Thinking it must be dirty, I wanted to pull bits of skin off, but that was still more painful. . . . I did not know I had been burnt. (Quoted in A. Scarth, *Savage Earth*, 1999)

Nearly all the forests around the Sunda Strait were incinerated. Recent research has shown that pyroclastic flows, being lighter, can actually skim across the surface of water; they are so hot and move so fast, lubricated by a carpet of steam beneath them, that they can even spread across ocean waters, from a volcano to more distant shores. For instance, Krakatau killed the 3,000 people then living on the island of Sebesi, about 13 kilometers (8 mi) away. There were many reports of human skeletons found floating on rafts of pumice all around the Indian Ocean; some of these would wash up in Africa nearly a year later.

The most deadly consequences, however, were the gigantic tsunamis. These were produced when the volcano collapsed into a huge, bowl-shaped caldera after the magma erupted—emptying the chamber that had held it and thus reducing the mountain's structural support—and sent shock waves through the surrounding water. These became enormous waves, similar to the tsunamis formed by earthquakes, which towered above the shoreline as they reached the shallow ocean bottom near the coast. The first explosion, of Perboewatan, sent a tsunami toward Telok Betong (now Bandar Lampung). The second explosion, which occurred under Danan at 6:44 a.m., sent huge tsunamis east and west. One of these walls of water, at least 46 meters (150 ft) high, destroyed the town of Merak. Most of the victims of Krakatau, in fact, were drowned by tsunamis. The official death toll was 36,417, but given the limited effort by Dutch officials to count people who had died among the native population, it might have been as high as 126,000.

The huge waves in the Sunda Strait were so powerful that they picked up coral blocks weighing as much 544 tonnes (600 short tons) and heaved them inland. The Dutch warship *Berouw* was carried 3 kilometers (2 mi) inland by a wave and stranded 10 meters (33 ft) above sea level on the tops of trees; all 28 of its crew members perished. There were so few survivors of the tsunamis that only a handful of eyewitness accounts are known. The following was recorded in the town of Anjer on Java:

> At first sight, it seemed like a low range of hills rising out of the water, but I knew there was nothing of the kind in that part of the Sunda Strait. A second glance—and a very hurried one it was—convinced me that it was a lofty ridge of water many feet high. . . . There was no time to give any warning, and so I turned and ran for my life. My running days have long gone by, but you may be sure that I did my best. In a few minutes, I heard the water with a loud roar break upon the shore. Everything was engulfed. Another glance around showed the houses being swept away, and the trees thrown down on every side. Breathless and exhausted I still pressed on. . . . A few yards more brought me to some rising ground, and here the torrent of water overtook me. I gave up all for lost. . . . I was soon taken off my feet and borne inland by the force of the resistless mass. I remember nothing more until a violent blow revived me. . . . I found myself clinging to a coconut palm. Most of the trees near the town were uprooted and thrown down for miles, but this one fortunately had escaped and myself with it. . . . As I clung to the palm-tree, wet and exhausted, there floated past the dead bodies of many a friend and neighbor. Only a mere handful of the population escaped. (Quoted in C. Officer and J. Page, *Tales of the Earth*, 1993)

Twelve hours later, tsunamis from Krakatau traveling at 480 kilometers (300 mi) per hour across the Indian Ocean reached the Gulf of Aden, south of the Arabian Peninsula. Ships as far away as South Africa were rocked violently as they lay in harbor. Smaller oscillations were reported in Japan, Australia, Hawaii, Alaska, and California as the waves worked their way out of the Indonesian archipelago and across the Pacific. There was even a disturbance in the tidal gauges of the English Channel.

By the next morning, August 28, Krakatau was eerily quiet. There were small mud eruptions but no further major activity. In place of the huge island with three major cones, Krakatau was now an enormous drowned caldera 250 meters (850 ft) deep. Three small islands were all that remained above water of the volcano's outer base (fig. 3.3). About 18 to 21 cubic kilometers (4.3 to 5.0 mi³) of volcanic material, mostly in the form of ignimbrites, or hot deposits formed by hot flowing ash and gases, had been displaced, settling over 1.1 million square kilometers (420,000 mi²). Some of this settled into the deep-ocean basins of the Sunda Strait around Krakatau or covered the shorelines of Java, Sumatra, and nearby islands.

Krakatau's effects were soon felt around the world. The enormous volume of ash that shot up to the stratosphere soon encircled the earth, blocking sunlight for well over a year. Global temperatures fell by 1° to 2°C (2° to 4°F) after the eruption and remained low for five years afterward. The ash and subsequent cooling also disrupted weather patterns, and the stratospheric dust scattered light in the orange and red parts of the spectrum. Unusual atmospheric events and abnormally vivid sunsets resulted, inspiring, as noted earlier in the book, artistic renditions of bright orange-red skies such as the one in Munch's *Scream* (1893). (Later the eruption would inspire a crummy 1969 Hollywood disaster film called *Krakatoa, East of Java*, which is so inaccurate that the title even gets the volcano's location wrong: it's west of Java.)

Krakatau remained quiet for a few more decades until December 27, 1927, when submarine eruptions were detected in the center of the caldera. A few days later, these eruptions built a new, much smaller volcano, called Anak Krakatau (Child of Krakatau), on top of the old vent. But even as ash and pumice were added to Anak Krakatau, the sea eroded them; not until 1930 did a series of lava flows build up the new island faster than the sea could wash it away. Anak Krakatau was growing at about 13 centimeters (5 in) per week when the current phase of eruptions began in the 1950s. Today it still goes through periods of rapid building, followed by several years of quiescence, followed in turn by renewed eruptions. The most recent significant eruption, in 2009, released hot gases, rocks, and lava in all directions.

The Dutch had ruled Indonesia for centuries, but the colony was in an advanced state of decay by 1883, with the Netherlands suppressing revolts by indigenous peoples in one war after another. In his 2005 book *Krakatoa: The Day the World Exploded*, Simon Winchester argues that the eruption of Krakatau helped to weaken the Dutch East Indies still further by crippling its economy and exposing the incompetence and corruption of its Dutch administrators. Although there had been unrest before, now the native Muslim population more actively revolted against their colonial overlords. By the time the Japanese invaded in January 1942, the Dutch colonial empire in Southeast Asia was virtually extinct. Yet even after Japan surrendered in 1945, the Netherlands tried to regain control of the islands. The attempt failed, and by 1949, Indonesia was independent.

Tambora and the Year without a Summer

The Krakatau eruption was huge, but it was dwarfed by an even bigger one that had happened 68 years earlier. When Mount Tambora, on the island of Sumbawa, east of Java, exploded in 1815, it became the largest eruption in recorded history (fig. 3.4).

The first sign of trouble came in 1812, when the volcano began to rumble, generating large earthquakes and a plume of ash. Without further warning, the main event started on April 5, 1815, with a thunderous explosion heard on Sulawesi, more than 380 kilometers (240 mi) away; in Batavia, 1,260 kilometers (780 mi) away; and in the Moluccan Islands, 1,400 kilometers (870 mi) away. Then, at about 7 p.m. on April 10, people on Sumatra, more than 2,000 kilometers (1,200 mi) away, thought they heard the sound of cannons, but it was actually the sound of an eruption. Three distinct columns of flame rose from the volcano and merged above it. The whole mountain turned into a liquid fire of pyroclastic flows, destroying the village of Tambora. April 11 brought more loud explosions, along with the odor of gases, and heavy ashfalls covered most of Indonesia. The sky stayed pitch black for two entire days as far as 600 kilometers (370 mi) away. The explosions also displaced the seafloor, creating tsunamis that hit adjacent islands with waves up to 4 meters (13 ft) high.

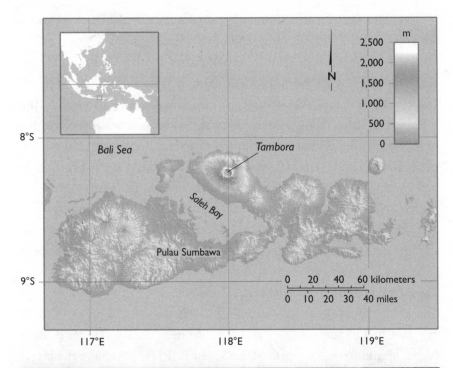

Figure 3.4. The eruption of Tambora. *a.* Index map showing the location of Tambora, seen here as the large volcano in the north-central part of Sumbawa. *b.* International Space Station image of the caldera as it looks today. (*a:* image by Sadalmelik; courtesy Wikimedia Commons; *b:* photograph by NASA Expedition 20 crew on the International Space Station; courtesy NASA Earth Observatory and Wikimedia Commons)

The biggest of the Tambora eruptions was thought to have a VEI rating of about 7 (with more than 100 km³, or 24 mi³, of ejecta), the largest in recorded history. In addition to hundreds of cubic kilometers of ash, which spread around the world, more than 41 cubic kilometers (9.8 mi³) of pyroclastic debris was ejected, which traveled at least 20 kilometers (12 mi) from the summit despite weighing about 10 billion tonnes (11 billion short tons). The formerly conical mountain had blown off its top, reducing its peak elevation from 4,300 meters (14,100 ft) to only 2,851 meters (9,354 ft) and creating a caldera about 7 kilometers (4 mi) across and 700 meters (2,300 ft) deep. Tambora killed between 70,000 and 90,000 people, 12,000 directly by means of the eruptions and the rest by starvation after the total destruction of crops in the region.

Sir Thomas Stamford Raffles, a local British lieutenant-governor, the founder of Singapore, and a respected naturalist, gave the most definitive description of the eruption and its aftermath in his 1830 *Memoir*:

Island of Sumbawa, 1815—In April, 1815, one of the most frightful eruptions recorded in history occurred in the mountain Tambora, in the island of Sumbawa. It began on the 5th day of April, and was most violent on the 11th and 12th, and did not entirely cease till July. The sound of the explosion was heard in Sumatra, at a distance of nine hundred and seventy geographical miles [1,700 km, or 1,000 mi] in a direct line, and at Ternate, in an opposite direction, at the distance of seven hundred and twenty miles [1,300 km, or 830 mi]. Out of a population of twelve thousand, only twenty-six individuals survived on the island. Violent whirlwinds carried up men, horses, cattle, and whatever else came within their influence, into the air, tore up the largest trees by the roots, and covered the whole sea with floating timber. Great tracts of land were covered by lava, several streams of which, issuing from the crater of the Tambora mountain, reached the sea. So heavy was the fall of ashes, that they broke into the Resident's house in Bima, forty miles [74 km, or 46 mi] east of the volcano, and rendered it, as well as many other dwellings in the town, uninhabitable. On the side of Java, the ashes were carried to the distance of three hundred miles [560 km, or 350 mi], and two hundred and seventeen [400 km, or 250 mi] towards Celebes, in sufficient quantity to darken the air. The floating cinders to the westward of Sumatra formed, on the 12th of April, a mass two feet [0.6 m] thick and several miles in extent, through which ships with difficulty forced their way. The darkness occasioned in the daytime by the ashes in Java was so profound, that nothing equal to it was ever witnessed in the darkest night. Although this volcanic dust, when it fell, was an impalpable powder, it was of considerable weight; when compressed, a pint of it weigh[ed] twelve ounces and three quarters [360 g]. Along the sea-coast of Sumbawa, and the adjacent isles, the sea rose suddenly to the height of from two to twelve feet [0.6 to 3.7 m], a great wave rushing up the estuaries, and then suddenly subsiding. Although the wind at Bima was still during the whole time, the sea rolled in upon the shore, and filled the lower parts of houses with water a foot [30 cm] deep. Every prow and boat was forced from the anchorage and driven on shore. The area over which tremulous noises and other volcanic effects extended was one thousand English miles [1,600 km] in circumference, including the whole of the Molucca Islands, Java, a considerable portion of Celebes, Sumatra and Borneo. In the island of Amboyna, in the same month and year, the ground opened, threw out water, and closed again.

Raffles sent Lieutenant Owen Phillips of the British Army to Sumbawa to observe the local conditions. Phillips reported back (as quoted in Raffles's *Memoir*):

On my trip towards the western part of the island, I passed through nearly the whole of Dompo and a considerable part of Bima. The extreme misery to which the inhabitants have been reduced is shocking to behold. There were still on the roadside the remains of several corpses, and the marks of where many others had been interred: the villages almost entirely deserted and the houses fallen down, the surviving inhabitants having dispersed in search of food. . . . Since the eruption, a violent diarrhoea has prevailed in Bima, Dompo, and

Sang'ir, which has carried off a great number of people. It is supposed by the natives to have been caused by drinking water which has been impregnated with ashes; and horses have also died, in great numbers, from a similar complaint.

Tambora had worldwide effects as well. It injected huge amounts of dust and gases, such as sulfur dioxide, into the stratosphere up to 43 kilometers (27 mi) above the planet's surface. Most of the coarser ash particles fell back to earth within a week or two, but ash and fine droplets of sulfuric acid remained in the stratosphere for several years, blocking sunlight. Thus, as the changing distribution of heat and cold transformed atmospheric currents, Tambora changed the planet's weather patterns.

As we learned in the prologue, the following year, 1816, came to be known as the "Year without a Summer" because that season was cold, dark, rainy, and even—in North America and Europe—snowy. Average global temperatures fell 0.7°C (1.3°F), which is a huge drop for the entire planet. People on the East Coast of North America first observed, in the spring and summer of 1816, a persistent "dry fog" that blocked sunlight and was unaffected by rains or wind. The fog was actually a fine mist of sulfates sitting in the stratospheric layer, about 8 kilometers (5 mi) above the earth's surface, and in the troposphere, the layer just below it, where we live. Nobody understood what the true cause of the weird weather was, but there was lots of speculation. Galileo had documented sunspots in his *Siderius Nuncius* (*Starry Messenger*) back in 1611, but this was the first chance for many people to see sunspots themselves, because the clouds of dust in the sky made the sun dim enough to look at directly. As Jeffrey Vail reports in his "'The Bright Sun Was Extinguish'd': The Bologna Prophecy and Byron's 'Darkness'" (1997), newspapers followed the phenomenon with interest. The *London Chronicle* wrote, "The large spots which may now be seen upon the sun's disk have given rise to ridiculous apprehensions and absurd predictions. These spots are said to be the cause of the remarkable and wet weather we have had this Summer; and the increase of these spots is represented to announce a general removal of heat from the globe, the extinction of nature, and the end of the world."

Even some scientists thought that the sudden visibility of these spots meant that the sun would soon go out. One prediction, claiming that the world would end on July 18, 1816, caused riots, suicides, and religious craziness all over Europe. Vail offers an account of one such incident:

> A Bath girl woke her aunt and shouted at her that the world was ending, and the woman promptly plunged into a coma. In Liege, a huge cloud in the shape of a mountain hovered over the town, causing alarm among the "old women" who expected the end of the world on the eighteenth. In Ghent, a regiment of cavalry passing through the town during a thunderstorm blew their trumpets, causing "three-fourths of the inhabitants" to rush forth and throw themselves on their knees in the streets, thinking they had heard the seventh trumpet.

The oddest part of 1816's weather was its extreme swings. Unlike in a normal seasonal pattern, when everything stays cold for months until spring comes, and nature and people grow accustomed to those conditions, the spring and summer of 1816 were marked by repeated freezes and thaws. Each time plants started to bud or crops were sown, they would enjoy a week or two of good weather and then be killed by a sudden freeze. Thomas Jefferson, by then an ex-president, had such severe crop failures at his home, Monticello, that he fell further into the debt that had plagued him throughout his life. Nicholas Bennet of the Church Family of Shakers in New Lebanon, in upstate New York, recorded in his journal in May 1816 that "all was froze" and the hills were "barren like winter" (quoted in Glendyne R. Wergland, *One Shaker Life*, 2006). Nearly every day that month, temperatures fell below freezing, and on June 9 the ground froze solid. Three days later, the Shakers replanted crops destroyed by the cold, only to lose them on July 7, when the ground froze again.

The *Columbian Register* of Norfolk, Virginia, complained on July 27, 1816: "It is now the middle of July, and we have not yet had what could properly be called summer. Easterly winds have prevailed for nearly three months past. . . . The sun during that time has generally been obscured and the sky overcast with clouds; the air has been damp and uncomfortable, and frequently so chilling as to render the fireside a desirable retreat." As far south as Pennsylvania, lakes and rivers froze even in July and August. In Massachusetts, the Berkshire Hills froze on August 23, as did much of the upper Northeast. William G. Atkins, a Massachusetts historian, wrote in his *History of the Town of Hawley* (1887):

> Severe frosts occurred every month; June 7th and 8th snow fell, and it was so cold that crops were cut down, even freezing the roots. . . . In the early Autumn when corn was in the milk it was so thoroughly frozen that it never ripened and was scarcely worth harvesting. Breadstuffs were scarce and prices high and the poorer class of people were often in straits for want of food. It must be remembered that the granaries of the great west had not then been opened to us by railroad communication, and people were obliged to rely upon their own resources or upon others in their immediate locality.

The lack of a true summer was not the only freakish weather of that year: the winter of 1816–17, too, was unusually harsh in North America and Europe, killing thousands of livestock. The overall set of poor conditions produced widespread disease, including a typhus epidemic in Ireland and the worst European famine of the 19th century. The failures of harvests in England and Ireland led to huge migrations of starving poor people; meanwhile, more than 100,000 died in Ireland alone and more than 200,000 in Europe as a whole. Many families in Wales became beggars and refugees as they ran out of provisions. In Germany, food prices skyrocketed, leading to demonstrations in front of markets and eventually riots, arson, and looting as starving people revolted against their ineffective leaders. Similar events transpired in many European countries, and the turmoil lasted as the effects of the disrupted climate patterns persisted into 1817 and even 1818. The poor weather couldn't have come at a worse time, as European governments were then trying to recover and stabilize after more than a decade of war that had ended with Napoleon's defeat at Waterloo in 1815.

The effects of the Year without a Summer were not restricted to Europe and North America. The climate change disrupted the monsoonal system in the Indian Ocean, leading to three failed harvests and widespread famine, with thousands of deaths in South Asia. A new strain of cholera appeared in the Bengal region in 1816 and soon killed thousands more. In China, the cold killed trees, crops, and even hardy water buffalo, especially in the colder northern regions. Even in Yunnan Province, in southern China, rice production was devastated as frost killed the crops and workers deserted their paddies. The Chinese monsoon system was also disrupted, producing huge floods in the Yangtze River valley. Even tropical Taiwan recorded snow and frost in the summer, rarely seen there even in winter.

The Year without a Summer triggered many cultural changes as well. The lack of oats and the widespread death of horses spurred the German inventor Karl Drais to research new methods of horseless transportation, giving rise to the ancestor of the modern bicycle. In the United States, the harsh climate in New England and New York drove many bankrupt farmers to lands farther west that had richer soils and were not affected by the cold summer of 1816. This region, then called the Northwest Territories, rapidly filled with people, so that Indiana became a state by December 1816 and Illinois by 1818. Meanwhile, Vermont lost 10,000 to 15,000 people, wiping out its previous seven years of population growth. One of the families that left the state then was that of Joseph Smith; they moved to Palmyra, New York, where Smith supposedly experienced the revelations that later yielded the Book of Mormon and the founding of Mormonism.

The chilly climate of the Year without a Summer and its corollary effects influenced even artists and writers. For years afterward, painters rendered landscapes with orange skies and spectacular sunsets, just as they had looked in 1816: J. M. W. Turner's circa 1828 *Chichester Canal* has a good example of such a yellow-orange sky. As the prologue mentions, Lord Byron and his guests, stuck inside his Swiss villa during that cold, dark, wet summer, passed their time by competing to tell the scariest gothic horror stories they could dream up. This contest eventually yielded Mary Shelley's *Frankenstein* (published in 1818), John William Polidori's *The Vampyre* (based on one of Byron's ideas and a predecessor of Bram Stoker's *Dracula*), and Byron's "Darkness," a gloomy and apocalyptic poem about the last man on earth. It was inspired, Byron wrote, by a day when "the fowls all went to roost at noon and candles had to be lit as at midnight."

Supervolcanoes

Although the Krakatau and Tambora eruptions were huge events, they are tiny compared to the enormous ones of the prehistoric past. Once a volcano has an eruption of VEI 8 or higher (producing more than 1,000 km³, or 240 mi³, of ejecta), it becomes known as a supervolcano. The geologist Edwin T. Hodge first proposed this term in 1925, arguing that a supervolcano that he named Mount Multnomah had collapsed into a caldera and thus formed the Three Sisters peaks in the Oregon Cascades. Hodge's theory eventually was disproved, and the debate over this purported volcano eventually subsided, as did usage of the term *supervolcano*. Then a BBC popular science show, *Horizon*, revived it in 2000 to describe immense eruptions that change the planet. In 2005, the BBC aired a two-part disaster film titled *Supervolcano*, and in 2006, the PBS science series *Nova* broadcast an episode called "Mystery of the Megavolcano." Normally, the scientific community ignores the hyped-up language of popular science TV shows and other media, but in this case it served a useful purpose among volcanologists. By the time of the 2006 meeting of the American Geophysical Union in San Francisco, the word was in such wide use that a special session was convened to talk about supereruptions, debate the criteria defining a supervolcano, and discuss whether the term is scientifically merited.

Eruptions on the scale of VEI 8 or higher are very rare, with only about a dozen documented in all of earth's history. One was the eruption of what is now La Garita Caldera in central Colorado, which blew its top 27.8 million years ago (see chapter 9). But the biggest event on the list in the past 27 million years was the eruption of Toba. As we learned in chapter 1, today Toba is a caldera on the north end of the island of Sumatra (fig. 1.3). When Toba erupted about 74,000 years ago, it ejected around 3,000 cubic kilometers (720 mi³) of material, second only to the eruption of La Garita. Toba produced about 2,000 cubic kilometers

(480 mi^3) of ignimbrites, which flowed across the ground, and more than 800 cubic kilometers (190 mi^3) of ash, which blew off to the west. (A more recent estimate suggests that Toba's eruption produced more than 3,200 km^3, or 1,200 mi^3, of ejecta.) Toba's pyroclastic flows destroyed about 20,000 square kilometers (7,700 mi^2), and the eruption left ash deposits as thick as 600 meters (2,000 ft) near the vent. An ash layer was spread all over Southeast Asia, about 15 centimeters (6 in) thick in most places and blown into drifts up to 6 meters (20 ft) deep at one site in central India. Malaysia was covered with ash deposits about 9 meters (30 ft) deep.

Geologists believe that Toba's main eruption took about two weeks. It is conjectured to have occurred during the Northern Hemisphere summer, since only then would the monsoonal conditions have blown ash as far north as the South China Sea, where Toba ash has been found. In addition to the solid material, it is estimated that Toba expelled about 10 billion tonnes (11 billion short tons) of sulfuric acid and 6 million tonnes (6.6 million short tons) of sulfur dioxide into the atmosphere, which had a huge effect on climate, as we will see later. As the prologue notes, Toba also released the energy equivalent of 1 million tonnes (1.1 million short tons) of TNT.

Toba had a long history of huge eruptions before the final big event. The geologist Craig Chesner (whom we met in chapter 1) has worked on the caldera for many years, mapping and dating all the eruptive deposits there. He has reported that about 1.2 million years ago, Toba's northern area erupted approximately 33 cubic kilometers (8 mi^3) of material, now known as the Haranggaol Dacite Tuff (HDT), in a VEI 6 event. The southern half of the present caldera (now called the Porsea Caldera) erupted approximately 800,000 years ago in an event that ejected roughly 500 cubic kilometers (120 mi^3) of material. The resulting deposit is known to geologists as the Old Toba Tuff (OTT). Then activity switched back to the northern part about 500,000 years ago, producing the Middle Toba Tuff (MTT) in a VEI 6 explosion that yielded 58 cubic kilometers (14 mi^3) of material.

The final Toba eruption, the one that this book focuses on, formed the Youngest Toba Tuff (YTT). It was by far Toba's largest eruption, with a VEI rating of 8 and a yield of 2,800 to 3,200 cubic kilometers (670 to 770 mi^3) of ejecta, as discussed above. Precise dating of the YTT eruption has been a challenge, since it is so old that error estimates (a result of the limitations of the equipment used to date the tuff) are large. The earliest scientific estimates dated it at about 75,000 ± 9,000 years ago—in other words, there is a 95 percent certainty that its true age lies between 84,000 and 66,000 years ago. This is the normal, expected error range for events that happened so long ago, but we would always like to date them more precisely so we can determine if they overlapped with specific climate signals or evolutionary changes.

The most recent dating of the YTT eruption was done in 2012 and 2013, using two independent methods. The first was argon-argon dating of mineral grains in the tuff, and it yielded ages of 73,880 ± 320 years and 75,000 ± 900 years, suggesting two separate eruptions. The second method looked at the sulfur chemical fallout from the ash clouds, found in both distant ice cores and in stalactites from caves in the Toba region. The long geologic time that they record has been calibrated precisely, using the astronomical cycles of the earth's orbit around the sun and its tilt on its axis, both of which affect climate. This method bracketed the YTT eruption between 74,450 and 74,050 years ago. That is incredibly precise for an event that happened tens of thousands of years ago, but currently there is no method of pinning down the age more precisely.

The YTT eruption injected so much ash and sulfuric acid into the stratosphere that it blocked the sun's radiation and caused global temperature to drop by 3° to 5°C (5° to 9°F), making subsequent ice age cycles colder than they would have been otherwise (chapter 7 further discusses the consequences of this event). Before we turn our attention to the effects such a temperature drop might have had on the earth's human population, however, we need to look at a completely independent source of evidence for the Toba eruption: data from molecular biology.

4 Clues in Your Genes

This [double helix] structure [of the DNA molecule] has novel features which are of considerable biological interest. . . . It has not escaped our notice that the specific pairing [of nucleotide bases] we have postulated immediately suggests a possible copying mechanism for the genetic material.

James Watson and Francis Crick,
"Molecular Structure of Nucleic Acids," 1953

The Blueprint of Life

In the early 1950s, scientists were in a race to discover the most important secret in all of biology: the chemical structure of the molecules that control inheritance. There were lots of ideas but no solid evidence. Biologists knew that the microscopic strands of material in the cell nucleus known as chromosomes must be the carriers of genes, but their precise makeup and how they worked was unknown. The conventional view was that genetic information was just a sequence of amino acids in proteins that could copy itself, although the mechanism for this was also uncertain. In the late 1940s, a number of labs had shown that chromosomes consist mostly of a molecule called deoxyribonucleic acid (DNA), whose existence had been known since 1869. However, at the time DNA was not believed to control inheritance. With its chemistry made up of just four nucleotides (organic molecules), it was considered to be a "stupid" tetranucleotide, its structure

69

too simple to play any role in inheritance beyond possibly supporting the copying of proteins by holding them together.

Yet important clues soon emerged suggesting that proteins are not the carriers of inheritance. In 1944, Oswald Avery, Colin MacLeod, and Maclyn McCarty of the Rockefeller Institute for Medical Research injected lab mice with DNA isolated from virulent pneumonial bacteria. They discovered that the bacteria reproduced (and killed the mice), thus demonstrating that DNA on its own carries the relevant instructions for reproduction. The researchers were sure they had proved that DNA was the carrier of inheritance, but most geneticists and molecular biologists ignored this result because they were still focused on proteins as the key to inheritance. In 1952, Alfred Hershey and Martha Chase at the Cold Spring Harbor Laboratory on Long Island showed that bacteriophages, viruses that infect bacteria, inject their DNA and not much else into host cells. However, they did not push hard for the obvious conclusion: that DNA must have something to do with inheritance in general and not just in bacteriophages. DNA was still dismissed as playing some as-yet undefined role.

But some were taking these early experiments further, especially an eager young geneticist named James Watson. Raised by a middle-class family on Chicago's South Side, he was precocious, even appearing on the early TV show *Quiz Kids*, which showcased gifted children. He enrolled at the University of Chicago at the tender age of 15, intending to become an ornithologist, as birdwatching was one of his favorite pastimes. Watson later described the university as an "idyllic academic institution where [I] was instilled with the capacity for critical thought and an ethical compulsion not to suffer fools who impeded [my] search for truth."

In 1946, while still an undergraduate, Watson read Erwin Schrödinger's classic 1944 book *What Is Life?*, which discusses the chemical basis for life. He realized that there were far more opportunities for groundbreaking research in the growing fields of genetics and molecular biology than he would find elsewhere. The following year, he earned his bachelor's degree and then entered graduate school at Indiana University to study with the legendary molecular biologists Hermann Muller and Salvador Luria.

Watson completed his doctorate in only three years, working in Luria's "Phage Group" lab, which focused on bacteriophages. In September 1950, he started a postdoctoral fellowship at Copenhagen University, working first with Herman Kalckar and then in Ole Maaløe's lab, which also studied phages. A year and a half later, in April 1952, he attended a Society of General Microbiology conference on the nature of viral multiplication. As he recounted in his best-selling 1968 book, *The Double Helix*:

> Several days before the meeting, Al Hershey had sent me a long letter from Cold Spring Harbor summarizing the recently completed experiments by which he and Martha Chase established that a key feature of the infection of a bacterium by a phage was the injection of the viral DNA into the host bacterium. Most important, very little protein entered the bacterium. Their experiment was thus a powerful new proof that DNA is the primary genetic material. Nonetheless, almost no one in the [conference] audience of over four hundred microbiologists seemed interested as I read long sections of Hershey's letter. Obvious exceptions were André Lwoff, Seymour Benzer, and Gunther Stent, all briefly over from Paris. They knew that Hershey's experiments were not trivial and that from then on everyone was going to place more emphasis on DNA. To most of the spectators, however, Hershey's name carried no weight.

Clearly, Watson and others could see that they were on the brink of finding out how the genetic code worked. Meanwhile, at Caltech, in the balmy climes of Pasadena, the legendary chemist Linus Pauling (one of the few people to win Nobel Prizes in two fields) was applying his genius to the problem. In 1951, he had published an analysis showing that the basic structure of a protein is a simple corkscrew shape known as an alpha helix. He continued to work with the concept of a helical structure for DNA. Unfortunately, he got sidetracked on the idea that DNA consisted of three strands wrapped in a triple helix and never returned to the right path.

Because of his controversial political views—he criticized the McCarthyism and "red hunting" of the period—the US State Department did not allow Pauling to travel abroad so he could find out what others were doing. However, Watson had read about Pauling's use of a new technique, X-ray crystallography, to decipher the structure of the alpha helix. X-ray crystallography shoots a beam of X-rays at the atoms in a crystal, and the angles at which it bounces back tell you the structure of the substance. Watson realized that this technique could be the key to discovering the structure of DNA and was eager to work on the problem. Luckily, his former advisor Salvador Luria had arranged the next step for him: in 1951, Luria had met John Kendrew of the now-famous Cavendish Laboratory at Cambridge University, and Kendrew invited Watson to do his next postdoc there.

The Old Cavendish Lab is one of the most important places in the annals of science. Founded in 1874 by the legendary physicist James Clerk Maxwell, who developed the mathematics of electricity and magnetism, it has hosted researchers who have won 29 Nobel Prizes, far more than any other institution in the world. It was the site of pioneering work in radioactivity by Ernest Rutherford, the discovery of the electron by J. J. Thomson, the discovery of the neutron by Arthur Compton, the isolation of argon by Lord Rayleigh and William Ramsay, and the development of X-ray crystallography by William Henry Bragg and his son Lawrence Bragg. The lab was among the first to synthesize plutonium and neptunium, which led to the controlled production of energy by nuclear fission. The breakthroughs at Old Cavendish continued for a century; in 1974, the overcrowded lab moved from its ancient downtown building to the outskirts of Cambridge.

After arriving in Cambridge, Watson soon befriended the physicist Francis Crick, who still had not finished his PhD despite being in his middle thirties. Crick did, however, have lots of experience reading the X-ray patterns of complex biomolecules. Together they set themselves to solving the puzzle of DNA's structure, struggling to get clear X-ray photographs and to find three-dimensional models that suited the data. They were aware that Pauling, too, was working on the problem and were worried that he would find the solution first. As Watson wrote in *The Double Helix*, "Our first principles told us that Pauling could not be the greatest of all chemists without realizing that DNA was the most golden of all molecules."

Then they learned that the lab led by John Randall at King's College London was also trying to obtain good X-ray images of DNA and other molecules. Several of the researchers there, especially Maurice Wilkins, Rosalind Franklin, and her student Raymond Gosling, had mastered the necessary techniques and were getting much clearer images than anyone else. In 1951, Watson attended a seminar at which Franklin presented some of her unpublished data, and in 1952, Crick read a Medical Council Research report that mentioned some of her work. Watson and Crick were soon collaborating with Wilkins, discussing the X-ray diffraction results extensively with him, but they didn't speak much to Franklin. They considered her as simply a junior lab researcher and not as the lead scientist. Wilkins never got along well with Franklin, either, since she was concise, impatient, and direct and looked people straight in the eye, while he was shy, slow to speak and react, and never made eye contact. A lot of people attribute the conflicts in the lab to Randall's lax management style, pointing out that he never dealt with the tension between the two.

Watson and Crick created physical "ball and stick" models of DNA molecules and their bonds (resembling Tinkertoys), manipulating them to see if the three-dimensional structure was feasible. One sticking point that hampered them was their insistence that the "backbone" of the DNA molecule ran down the middle, like an animal's spine, and that the other parts hung off it like Christmas tree ornaments. (Pauling was working under the same misconception.) But in a critical meeting in 1952, Franklin showed them that the backbone had to be on the outside of the molecule, and this completely changed the way they constructed their models.

By January 1953, Watson and Crick were convinced that the structure of the DNA molecule was a double helix, with the two strands of the external backbone spiraling around each other. On January 30, Watson went to King's College with an unpublished draft of Pauling's idea of the triple helix structure. Wilkins was not in his office or his lab, so Watson spoke to the other people in the lab, urging them to collaborate with Crick and him before Pauling discovered his mistake. Franklin grew angry at the implication that she did not understand her own data and lashed out at Watson, who backed up and bumped into Wilkins, then returning to the lab to investigate the source of the ruckus. Wilkins tried to console Watson privately in his office and showed him Franklin's best image (known as Photograph 51), while Watson showed Wilkins his copy of Pauling's manuscript.

Photograph 51, which Franklin had never published, was the evidence that Watson and Crick were seeking. Wilkins made it available to them as the lead scientist in the King's College lab, which Franklin would soon leave to move to Birkbeck College. By convention, all her lab work at King's was property of the university and had to be left behind.

On February 28, 1953, Watson and Crick began building a physical model that worked. When they realized they had it right, they rushed down the street to the Eagle Pub, where Old Cavendish researchers frequently ate lunch. As Watson described it in *The Double Helix*, "Francis winged into the Eagle to tell everyone within hearing distance that we had found the secret of life." They finished their model a few days later. The following day, they received a letter from Wilkins saying that Franklin was leaving at last and that they could go ahead and use the data from the King's College lab: in Wilkins's words, they could put "all hands to the pump." Over the next few weeks, Wilkins went up to Cambridge, saw Watson and Crick's model, and brought the word to Gosling when he returned to London. Meanwhile, Watson and Crick rushed to write up their results before Pauling could scoop them. On April 2, 1953, they submitted their paper to the British journal *Nature*, and on April 25, their paper was published and changed biology forever.

Meanwhile, what happened to Franklin, whose work was part of the foundation of Watson and Crick's models? One problem was that for much of 1952, she was sidetracked by asymmetrical X-ray patterns, coming to recognize DNA's helical structure only with better images that she obtained in early 1953. She was also was a much more cautious scientist than Watson or Crick, believing that she needed to gather more evidence of the structure of DNA before publishing her results. Franklin was not a big believer in molecular models, either, although photos of her Birkbeck lab show that she did use small ones.

Despite her unhurried, careful stance toward publishing, she did try to get credit for her work. She wrote a draft of her interpretation of DNA's double helix structure and submitted it to *Acta Crystallographica* on March 6, 1953, but it was not published until late summer. Because that journal was not nearly as prominent or as fast to publish as *Nature*, one of the highest-profile and speediest journals in all of science (along with the American *Science*), her work was beaten to print. Even after Watson and Crick's paper came out and she saw their model at the Old Cavendish Lab, she said to others there, "It's very pretty, but how are they going to prove it?" Ever the cautious scientist, she was still not convinced that her own work was conclusively proved and was certainly not ready to put her name on the idea that Watson, Crick, and Wilkins made world famous.

Meanwhile, the famous crystallographer J. D. Bernal recruited Franklin to Birkbeck because he could see the importance of her work; he also had a penchant for hiring brilliant but neglected women scientists. Her move, planned for a long time but delayed until March 1953 because of previous contractual commitments, was driven by her desire both to get away from the tense conditions in the King's College lab and to become the senior scientist of her own research group, where no one could steal her work or credit. However, by changing institutes, she also sacrificed and moved down the ladder of academic and research prestige to a small college with limited resources, "moving from a palace to the slums . . . but pleasanter all the same," as she put it, according to Brenda Maddox's 2003 biography.

Once at Birkbeck, Franklin worked on publishing what she had discovered. Her article on the alpha helix structure of DNA appeared in *Nature* on July 25, 1953, and gained her higher-profile recognition of her work, whose importance Watson and Crick had minimized in their own paper. Then she moved on to study the structure of the ribonucleic acid (RNA) molecule, working with the tobacco mosaic virus to decipher how it transferred its RNA to a host organism. Bit by bit, she obtained more recognition, and she solved many complex problems in molecular biology. In fact, her model of the tobacco mosaic virus was displayed at the International Science Pavilion at the Brussels World's Fair on April 17, 1958. However, Franklin had died the day before of bronchopneumonia and ovarian cancer, possibly brought on by her frequent exposure to X-rays. She was only 37.

Rosalind Franklin's place in the history of science is a hotly debated topic. The sexism of the King's College lab is well documented. Clearly, the lab prevented women from getting their fair share of resources or attention, which was part of the reason that Franklin accepted the lesser post at Birkbeck College. In addition, the lab followed the horrendous practice, still standard today, of making junior researchers give up all their results when they leave a lab and allowing their supervisors to take all the credit. Most scholars agree that Franklin should have received more attention for her X-ray images, which made Watson and Crick's discovery possible. They were remiss in not sufficiently acknowledging her work—Watson even minimizes and dismisses its importance in *The Double Helix*—but evidence shows that Franklin was also much more cautious and reluctant to publish, and she never would have agreed to coauthor with Watson and Crick because of their difference in attitude. They were much more willing to gamble with the limited data they had and go for the big score. Franklin's eventual publication of her work came too late to garner her a fair share of all the attention that Watson and Crick received by publishing first, and in the highest-profile journal. In addition, the scientific community continued to remain cautious about whether Watson and Crick's double helix model was correct at all, and it wasn't until 1961 that DNA's identity as the secret of inheritance was finally proved (see the next section).

Still, in the crowning injustice, Watson, Crick, and Wilkins received the Nobel Prize in Physiology or Medicine in 1962, not only for their discovery of the DNA double helix but also for their continued work on the structure of DNA and RNA and for deciphering the genetic code. Even though she deserved it, Franklin could not have shared in that prize: under the Nobel Foundation's rules, the award can never be given posthumously and cannot be split more than three ways. To add insult to injury, Franklin's protégé Aaron Klug received the 1982 Nobel Prize in Chemistry for his research on the X-ray crystallography of nucleic acids. This was work she pioneered, and she probably would have shared that Nobel had she lived that long.

Too Many Genes

After the first flush of research suggesting that DNA is the blueprint of life and the ensuing decipherment of its detailed structure, biologists worked to understand its copying and translation mechanism and then moved on to an even larger topic: finding the genetic code itself.

Early studies of DNA showed that the double helix consists of two external backbones made of phosphates and sugars (such as ribose or deoxyribose), arranged in a spirally twisted ladder whose rungs are a series of nitrogenous bases known as adenine, thymine, guanine, and cytosine (A, T, G, and C; in RNA, uracil, U, replaces thymine). Combined with the sugar and phosphate of the ladder, this unit of bases, sugars, and phosphates is known as a nucleotide.

As early as 1959, a group led by Crick found that a sequence of three nucleotides (called a codon) is all that is necessary to code for a protein. Then Marshall Nirenberg and Heinrich Matthaei of the National Institutes of Health (NIH) used a clever technique to figure out which codon of three nucleotides specifies which protein. They synthesized an RNA strand of nothing but uracil (UUUUU . . .), which produced only the protein phenylalanine. Nirenberg presented their results at the International Congress of Biochemistry in Moscow in 1961, and Crick was so impressed that he persuaded the entire congress to listen to Nirenberg's talk again the next day. Nirenberg and Matthaei then showed that a sequence of nothing but adenine (AAAAA . . .) produces the protein lysine, while nothing but cytosine (CCCCC . . .) produces the protein proline. Soon all of molecular biology was focused on the "coding race," with several labs competing to see who could decipher the genetic code first.

Severo Ochoa's lab at New York University, with its large staff, was in the lead. Nirenberg's small NIH lab could not compete, so many NIH scientists laid down their own research to help Ochoa's researchers sequence as much of the code as possible. DeWitt Stetten, the lab's director, later called the joint effort "NIH's finest hour." Finally, later in the early 1960s, Har Gobind Khorana, of the University of Wisconsin, Madison, deciphered most of the remaining parts of the code. The finding of the genetic code was so momentous that in 1968, Nirenberg, Khorana, and R. W. Holley (who discovered transfer RNA, which reads the genetic code) shared the Nobel Prize in Physiology or Medicine.

As the genetic code was deciphered, scientists were shocked by its redundancy. Of the three-base sequences, all that usually matters is the first two "letters" in the code (fig. 4.1). The third base is typically redundant and does not change which amino acid is produced. For example, any sequence that begins with GU produces valine, one that begins with AC produces threonine, and a sequence starting with CG produces arginine. Only a few of the codes require the third letter to specify which amino acid they produce, and even then there are usually two options (for example, CAU and CAC both produce histidine, while AAA and AAG both produce lysine). The 64 possible three-letter combinations of the four bases specify only the 20 amino acids that life uses, plus a "start" code, which begins transcription, and some "stop" codes, which end it.

Many scientists realized that mutations in the third, "silent" position in the codon are invisible to natural selection. The idea that much of the genetic material is apparently not affected by natural selection and is therefore selectively neutral was developed at length by Motoo Kimura in a 1968 paper, "Evolutionary Rates at the Molecular Level." J. L. King and Thomas Jukes made an even more radical argument in their 1969 paper "Non-Darwinian Evolution," stating that *most* mutations cannot be detected and are thus selectively neutral. Working independently, Kimura, King, and Jukes helped to develop what became known as the neutral theory of evolution.

This theory shocked the community of evolutionary biologists. Most were wedded to the notion of panselectionism, which holds that every

2nd base

		U		C		A		G		
	U	UUU	Phenylalanine	UCU	Serine	UAU	Tyrosine	UGU	Cysteine	U	
		UUC	Phenylalanine	UCC	Serine	UAC	Tyrosine	UGC	Cysteine	C	
		UUA	Leucine	UCA	Serine	UAA	Stop	UGA	Stop	A	
		UUG	Leucine	UCG	Serine	UAG	Stop	UUG	Tryptophan	G	
1st base	C	CUU	Leucine	CCU	Proline	CAU	Histidine	CGU	Arginine	U	3rd base
		CUC	Leucine	CCC	Proline	CAC	Histidine	CGC	Arginine	C	
		CUA	Leucine	CCA	Proline	CAA	Glutamine	CGA	Arginine	A	
		CUG	Leucine	CCG	Proline	CAG	Glutamine	CGG	Arginine	G	
	A	AUU	Isoleucine	ACU	Threonine	AAU	Asparagine	AGU	Serine	U	
		AUC	Isoleucine	ACC	Threonine	AAC	Asparagine	AGC	Serine	C	
		AUA	Isoleucine	ACA	Threonine	AAA	Lysine	AGA	Arginine	A	
		AUG	Methionine (start)	ACG	Threonine	AAG	Lysine	AGG	Arginine	G	
	G	GUU	Valine	GCU	Alanine	GAU	Aspartic acid	GGU	Glycine	U	
		GUC	Valine	GCC	Alanine	GAC	Aspartic acid	GGC	Glycine	C	
		GUA	Valine	GCA	Alanine	GAA	Glutamic acid	GGA	Glycine	A	
		GUG	Valine	GCG	Alanine	GAG	Glutamic acid	GGG	Glycine	G	

Figure 4.1. The three-letter codon sequences that dictate which amino acid is produced in making proteins. (National Institutes of Health; courtesy Wikimedia Commons)

variation in DNA, no matter how slight, is under the control of natural selection, whether we can detect how the selection works or not. This idea goes back to Charles Darwin, who wrote in *On the Origin of Species* (1859): "It may be said that natural selection is daily and hourly scrutinising, throughout the world, every variation, even the slightest; rejecting that which is bad, preserving and adding up all that is good; silently and insensibly working, whenever and wherever opportunity offers, at the improvement of each organic being in relation to its organic and inorganic conditions of life."

I vividly remember taking evolutionary biology courses from hardcore neo-Darwinists at Columbia University in the late 1970s, and this kind of strict panselectionism was still dogma to them. But at the very time when I was being taught this outdated notion, the neutral theory of evolution was becoming better and better established. It was clear that the third position in a codon is nearly always silent and thus that any random mutation in this position will have no effect on the resulting protein and so will be invisible to selection. Therefore, almost a third of the genetic code is completely neutral and cannot be seen or affected by external selection.

The redundancy of the genome became even more apparent when a 1966 experiment by Richard Lewontin and Jack Hubby showed that most organisms have far more genetic material than they need to produce a functioning organism; much of the extra is noncoding DNA (nicknamed "junk DNA"), which is unread by the molecules that transcribe DNA. In the 1970s and 1980s, molecular biology matured enough to be able to decipher amino acid sequences, and by 2000, the complete DNA of humans had been sequenced.

The most striking evidence supporting the idea that most DNA is adaptively neutral is the fact that the size of a genome often bears no relation to the complexity of the organism. This enigma was nicknamed "the onion test" by the Canadian biologist T. Ryan Gregory, referring to the surprising fact that a common onion has five times more DNA than a human yet is much simpler. Some salamanders have 35 times as much DNA as humans, and lungfish have 40 times as much DNA as we do. One species of deer has 20 percent more DNA than a close relative does, and one species of puffer fish has 100 times as much DNA as another. Among plants, there is no correlation between complexity and DNA: the broad bean has four times the DNA of a kidney bean, for instance. Even some single-celled microbes have more DNA than humans, while simple roundworms (nematodes) and watercress have about the same amount of DNA as a human. At best, there is only a rough correlation between DNA amount and complexity: single-celled organisms often do have smaller genomes than complex organisms (fig. 4.2). But if an onion has five times as much DNA as humans do, and lungfish 40 times as much, clearly most of that extra is not coding for more structures.

It's even possible to delete some of the repetitive noncoding sequence and produce viable animals. In 2004, an experiment deleted almost 3 percent of the genome of mice, which then reproduced with no ill effects. Again, if this DNA were functional, how could the mice produce fit offspring without it?

What is this seemingly useless DNA doing? Some of it may maintain the spacing between coding regions or help to hold the shapes of complex

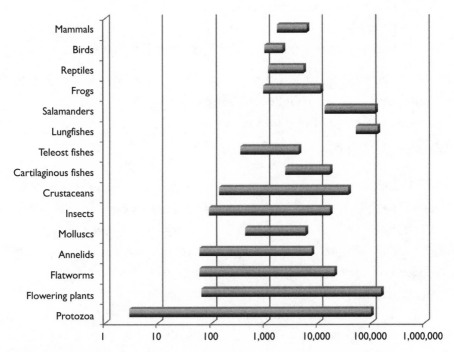

Figure 4.2. Amount of DNA in various groups of organisms. Clearly, there is no simple relationship between an organism's amount of DNA and its complexity. (Courtesy T. Ryan Gregory)

folds of long DNA strands. Segments of noncoding regions include the following:

- introns, chunks of DNA that are initially read but then edited out during final gene splicing;
- pseudogenes, chunks of DNA that have lost their ability to code for proteins;
- repetitive DNA, the same codons repeated over and over again hundreds of times and apparently coding for nothing;
- transposons, or "jumping genes," which can move from one part of the DNA to another yet are not expressed;
- short interspersed nuclear elements (SINEs) and long interspersed nuclear elements (LINEs), chunks of DNA stuck in the middle of a coding sequence that have no function or ability to code for proteins; and

- highly conserved noncoding nonessential DNA, which is very consistent in the sequences of many organisms, suggesting that it is important, yet can be removed with no effect whatsoever.

Perhaps the most interesting and surprising of the many kinds of junk sequences are endogenous retroviruses (ERVs). These are gene sequences of viruses that infected organisms long ago by inserting their DNA into our genome. Although this "fossil DNA" is no longer active, every time one of our cells divides, we make new copies of it. Because the ERVs hiding in our genome no longer code for the original viruses—or for anything else—they are clearly junk that we passively carry around with no ill effects.

Panselectionist biologists have had trouble addressing all this evidence and instead cling to bits of biology that seem to support their belief system. In 2012, for instance, the media made a big fuss when the Encyclopedia of DNA Elements (ENCODE) research project argued that maybe 80 percent of the human genome codes for some kind of protein. Naturally, many panselectionists took this to confirm their belief that *all* of human DNA is functional. So although ENCODE still concluded that at least 20 percent of our DNA is noncoding, panselectionists ignored this and proclaimed that they had been vindicated.

However, it turned out that even the ENCODE results were too good to be true. A study by Dan Graur and colleagues in 2013 demolished the ENCODE assertions, reaffirming that most of the genome (at least 90 and perhaps as much as 98 percent) is indeed noncoding. The ENCODE study managed to show only that some of the genome called "junk" does code for proteins. What it didn't show is that these random, isolated proteins are part of a functional biochemical pathway or that they lead to any visible consequences. If a protein results from junk DNA but doesn't do anything, it's still junk.

Clockwork DNA

The fact that most of the DNA in any organism is selectively neutral and apparently codes for nothing explains another discovery made in the

1960s. In 1962, Linus Pauling and Émile Zuckerkandl noticed that the number of genetic differences between two related species is proportional to how long ago they split apart. In other words, the longer two lineages evolve in two different species, the more genetic differences they accumulate. In the 1960s, we still could not sequence the complete DNA of organisms, but we could glean information about genetic similarity between species from the protein sequence in much simpler biomolecules, such as cytochrome c, which transfers electrons with the mitochondrion. As another molecular biologist, Emanuel Margoliash, wrote in his 1963 paper "Primary Structure and Evolution of Cytochrome C":

> It appears that the number of residue differences between the cytochrome c of any two species is mostly conditioned by the time elapsed since the lines of evolution leading to these two species originally diverged. If this is correct, the cytochromes c of all mammals should be equally different from the cytochromes c of all birds. Since fish diverges from the main stem of vertebrate evolution earlier than either birds or mammals, the cytochrome c of both mammals and birds should be equally different from the cytochromes c of fish. Similarly, all vertebrate cytochromes c should be equally different from the yeast protein.

This remarkable discovery suggests that DNA changes constantly over time. Even more surprising, the number of such differences between species is highly consistent with their time of divergence, so, for instance, most vertebrates differ by only 13 to 14 percent in molecules such as cytochrome c, while the cytochromes c of more distant organisms, such as plants and yeast, differ by 64 to 69 percent. This roughly constant rate of change, like the ticking of a watch, has come to be known as the molecular clock. As the method for estimating genetic distance by looking at simple biomolecules developed, it allowed molecular biologists to estimate when various lineages branched off from one another in the evolutionary tree of life (fig. 4.3).

Constant, regular changes in DNA could not slowly accumulate through random genetic accidents and genetic drift unless they were

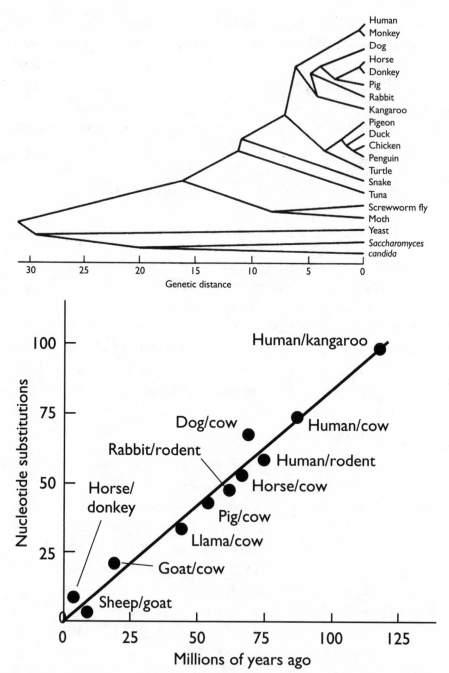

Figure 4.3. *a.* Family tree of the cytochrome *c* molecule, showing a branching pattern that is virtually identical to the sequence of branching events in the history of life. *b.* Linear relationship between the amount of genetic change (nucleotide substitutions) and the time that species diverged in their evolution millions of years ago. (Both redrawn from several sources)

invisible to natural selection. If most of the DNA were under the strict control of natural selection, there would not be such a tight match between divergence time and genetic distance, because selection would speed up or slow down the rate of mutation. The fact that changes in the genetic code tick away in a clocklike fashion, unaffected by natural selection, adds to the large pile of evidence supporting neutralism, including the silent third position in most codons and the presence of so much junk DNA in most genomes. Our current picture of genes has changed a lot since Watson and Crick's discovery in the 1950s; today we know that most DNA is unread and neutral, or invisible, to natural selection.

"Mitochondrial Eve"

Once the concepts of neutralism, junk DNA, and the molecular clock had been established in the 1960s and 1970s, increasing numbers of analyses used molecular clock methods to determine the time of divergence of different lineages. These studies often focused on the DNA of the cell nucleus, which has a relatively slow mutation rate. Then another surprising discovery was made: the mitochondria, the "power plants" of the cells of all animals, which produce energy, have their own, separate DNA. Later, it was established that this DNA is a relic of the days when mitochondria were free-living purple nonsulfur bacteria, which first lived symbiotically within the cells of animals and eventually developed into organelles in the cells. Even more interesting, not only is mitochondrial DNA different from that of the nucleus, but its molecular clock ticks five to 10 times faster, with about 0.02 substitutions per base (or about 1 percent of the total DNA) occurring every million years. For this reason, it can be used to examine genetic divergences that have happened in the past few thousand to few hundred thousand years, in contrast to the millions of years necessary for similar changes in nuclear DNA.

Using this fact, in 1980 Wesley Brown looked at the mitochondrial DNA of 21 women and found that they all had a common ancestor who lived about 180,000 years ago. Only women could participate in this study because mitochondria are carried in the protoplasm of women's ova, not in men's sperm. Allan Wilson at the University of California,

Berkeley, saw the potential to develop a DNA "clock" for humans from this line of research, so he and his graduate student Rebecca Cann sampled DNA from as wide a variety of people of different races as possible. They focused especially on getting samples from people who belong to one of the huge number of distinct populations in Africa—the continent that, given the fossils that have been collected there since the 1960s, is known to be the homeland of humanity. As the sample size increased, the estimates of the divergence times of human lineages got better and better, although there were still some major uncertainties and problems. Wilson, Cann, and Mark Stoneking (another graduate student in Wilson's lab) ended up with 147 samples from women all over the world and submitted a paper to *Nature* in late 1985. It was so controversial that the journal took more than a year to review and publish it, during which time it underwent some 40 revisions (normally a hot idea speeds through this process in a few months). It finally appeared on New Year's Day 1987.

What Cann, Wilson, and Stoneking had discovered was startling. Their samples represented every known genetic grouping on earth, with the majority from Africa. The molecular tree they produced showed that all humans came from Africa, since the most ancestral DNA sequences in their samples were all African. Even more surprising was the fact that all human mitochondrial DNA comes from a common ancestor: a woman who lived in Africa between 140,000 and 200,000 years ago (which is consistent with Brown's original estimate). In other words, most human fossils more than 200,000 years old are not closely related to anyone living today but instead belong to members of extinct genetic lineages. All living humans descend from populations that arose in Africa and evolved there through most of human prehistory. Only late in human evolution did the ancestors of today's humans migrate to other continents. This finding was a fatal blow to the now-discredited multiregional hypothesis, which argued that different races evolved in parallel on different continents from lineages established 500,000 to more than a million years ago. In this model, modern Asians would descend from Chinese or Javan *Homo erectus* (see chapter 6), while modern Europeans and modern

Africans would likewise have evolved independently. Instead, in reality, all modern humans descend from African ancestors who left that continent relatively recently, less than 100,000 years ago.

If the idea of a common female ancestor for everyone on earth was not controversial enough, the media managed to confuse things even further. Roger Lewin, who covered anthropology for *Science*, titled his write-up in that journal "The Unmasking of Mitochondrial Eve," while the issue of *Nature* with the Cann, Stoneking, and Wilson paper accompanied it with a summary titled "Out of the Garden of Eden." This led to a wide set of misconceptions, and Cann and her coauthors had to waste a lot of time clearing them up before they could talk about what their paper really says. First, the misleading "Mitochondrial Eve" and "Garden of Eden" metaphors that science writers—not the paper's authors—used somehow gave creationists the idea that this study had found the DNA of the biblical Eve herself, or had confirmed that the Garden of Eden was real, or, even worse, had disproved evolution. Apparently not enough people read past the headlines to reach the key part of the research, which established the geological age of humans' oldest common DNA ancestor at 140,000 to 200,000 years ago, far past the 6,000-year time frame to which young-earth creationists subscribe.

If the "Eve" who lived back then was typical of humans who leave fertile living descendants, she would have had numerous children and grandchildren, dozens of great-grandchildren, hundreds of great-great-grandchildren, and so on. This is simply a function of how reproduction, exponential growth, and family trees work. After about 140,000 to 200,000 years, her descendants would have made up a significant portion of all humans on earth, provided a population crash didn't prune them back. All the other females who were alive at the same time as this "Eve" may also have left lots of descendants, but for some reason none of them survive today, at least among the 147 women around the world whom Cann, Wilson, and Stoneking sampled.

Despite all the fire and fury that their paper provoked, it has held up under the scrutiny of a number of independent labs for more than 30 years. Most recent estimates confirm that all humans descend from a single African female with an age within the limits suggested in 1987. For example, one 2013 study gave the divergence time of the common DNA ancestor as 160,000 years ago (in the middle of the original range), while another study from 2013 produced the slightly younger range of 99,000 to 148,000 years ago. Thus, the original conclusion, including the dating, is robust and well established.

In a parallel study in 1995, Michael Hammer looked at genetic similarities among Y chromosomes in males all over the world. He found that all living humans descend from a common African father who lived about 200,000 to 300,000 years ago, which is consistent with the appearance of the first anatomically modern human fossils in Africa (see chapter 6). Naturally, the media mislabeled this ancestor the "Y-chromosome Adam," spreading the same misconceptions all over again. When the Y chromosome of a Neanderthal was sequenced in 2016, it placed the divergence between Neanderthals and *Homo sapiens* at 588,000 years ago, again consistent with the age of the divergence shown in the human fossil record.

The next question is when human populations left Africa. The latest data suggest that the "Out of Africa I" expansion occurred between 400,000 and 200,000 years ago, resulting in the Neanderthals, the Denisovans, and possibly some *Homo sapiens* in Eurasia (see chapter 6). However, several lines of genetic evidence show that none of these migrating humans have living descendants; all their lineages died out at one time or another.

According to the genetic evidence for the "Out of Africa II" expansion, the ancestors of native peoples in Eurasia and the Americas did not leave Africa until about 71,000 to 77,000 years ago. Apparently, they migrated along the southern coastline of Asia and colonized Australia about 50,000 years ago. One lineage of this migration arrived in southern Europe about 55,000 years ago from the Middle East (which had had modern humans some time earlier, probably 200,000 years ago, but one find in North Africa suggests possibly 300,000 years ago). Finally, the lineages that crossed the Bering land bridge reached the Americas only about 13,000 to 16,000 years ago, although some controversial dates from the Monte Verde site in Chile suggest that they arrived possibly as soon as 30,000 years ago.

Through the Bottleneck

Molecular biology made enormous strides throughout the 1990s and 2000s. The biggest breakthrough was the discovery of the polymerase chain reaction (PCR), which made it possible to sequence huge amounts of DNA in a matter of days or weeks rather than months or years. In the late 1990s, the complete genomes of several organisms were deciphered. The year 2000 brought not only the end of a millennium but also the sequencing of the human genome. This amazing feat was accomplished by both Craig Venter's group at Celera Genomics and the huge international Human Genome Project, led by Francis Collins (who took over from James Watson).

Given all this work, we now have the full DNA sequences of a number of organisms, including fruit flies; lab rats, mice, and rabbits; several domesticated animals; the nematode worm *Caenorhabditis elegans*; and most of our ape relatives: chimps, gorillas, orangutans, and gibbons. The mitochondrial DNA of some apes was sequenced as early as 1982, but that of chimpanzees was not sequenced until August 2005. The DNA sequences of more organisms are completed every year, and even better, they're all publicly available in a number of online genetics databases with easy, free access. With data sets such as these, it's possible to make many discoveries that were simply impossible before. Using molecular distances between diverging lineages, researchers have mapped most of the tree of life and solved many difficult problems (including the question of how all the major groups of mammals are interrelated).

In addition to genetic similarity and distance, there's another interesting thing that can be read in genomes: population size. If you compare the genomes of many individuals of the same species, you will see that some populations have a lot of genetic variability, while others have very little. This is a well-known consequence of a crash in population size, known as a genetic bottleneck. Because a species recovers from such an event from a small number of survivors, its later populations not only have reduced genetic diversity but might also have high frequencies of some genes that were originally rare in the species.

There are many examples of extant species that have passed through bottlenecks. The European bison, or wisent (*Bison bonasus*), was nearly wiped out, with only 12 individuals left in the early 20th century. Today all living wisents are descendants of those 12 survivors and have very low genetic diversity. Likewise, overhunting of the American bison (*Bison bison*) reduced a population that numbered approximately 60 million before Columbus arrived in the New World in 1492 to about 750 individuals, living in a few protected reserves, by 1890. Now they have recovered to more than 360,000 animals, including large populations in parks such as Yellowstone in Wyoming, Montana, and Idaho and Custer State Park in the Black Hills of South Dakota, along with many that are privately ranched for their meat—but they too are very low in genetic diversity. The northern elephant seal was down to 30 individuals in the 1890s, thanks to overhunting, but there are now hundreds of thousands—with very low genetic diversity. Cheetahs are famous in biology for having extremely low genetic diversity, probably because of a bottleneck in their population about 10,000 years ago, when most large mammals died out at the end of the last ice age. Even the familiar golden hamsters that are so common in pet stores are descendants of a single litter found in the Syrian desert in the 1930s, which became the ancestors of nearly all living hamsters—while the wild population from which they were taken is now nearly extinct.

Living and breeding with such low diversity can be a problem for many species. Inbreeding often leads to rare bad genes propagating throughout descendant populations. In addition, if a pathogen spreads to a genetically uniform species without the genetic instructions to make the antibodies to fight it off, the population might be wiped out. Low-diversity populations can thus be sitting ducks, vulnerable to any rapid change in their environment that they don't have the genes to adapt to.

Bottlenecks in Humans

Humans too apparently passed through a bottleneck. It's shocking, given that the human population on earth today is 7.6 billion and rapidly increasing, to realize how small that number was in prehistory.

Modern speciation theory suggests that most species are born of a small subpopulation that is isolated from the main population and no longer interbreeds with it (or other subpopulations). When such a sub-population shrinks, if its members have a large enough frequency of unusual genes, they can become the founders of a new species with a new genetic makeup. This "founder effect" is thought to have occurred at least once among the australopithecines (our ancestors from Africa, which lived between 4 and 2 million years ago), although we have no australopithecine genomes left and so cannot evaluate this claim (see chapter 6). In 2005, a group of scientists at Rutgers University found that genetic evidence suggests that all the Native American populations in the Americas arose from only about 70 individuals, who spread there from Asia about 11,000 to 13,000 years ago. This means that all the origi-nal peoples of the Americas—from the Inuit of Alaska to the Lakota of the Plains to the Aztecs, Olmecs, Toltecs, and Maya of Mesoamerica to the Incas of western South America and even the Fuegians of Patagonia—are extremely closely related and low in genetic diversity.

Studies of the full diversity of living *Homo sapiens* show that our species went through a very narrow bottleneck not that long ago. The most recent genetic evidence places the total population in the bot-tleneck at 30,000 people, although one estimate has placed it as low as 40 breeding pairs. Most hypotheses fall between 4,500 and 5,000 individuals, which is less than the population of the average small town in America.

The low genetic diversity of this population has important implica-tions. For centuries, the scientific community has been deeply racist and treated nonwhite peoples as inferior or even as different species from white people. But our molecules paint a completely different picture. When you compare the DNA of all the human "races" to the DNA of Neanderthals and that of most living species of chimps and gorillas, a

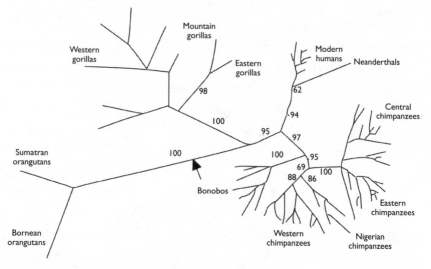

Figure 4.4. Branching tree of relationships of various populations of apes and humans, revealing their genetic distance from one another based on mitochondrial DNA. (Numbers indicate total genetic distance between different populations.) All human "races" are much more similar to one another than any two populations of gorillas or chimpanzees are to each other. (Modified from Pascal Gagneux et al., "Mitochondrial Sequences Show Diverse Evolutionary Histories of African Hominoids," *Proceedings of the National Academy of Sciences USA* 93 [1999]: fig. 1b, © 1999 National Academy of Sciences USA)

surprising result emerges. As fig. 4.4 shows, the genetic differences among the human races are extremely slight. All humans are far more genetically similar to one another than the populations of West African chimps are to each other, and the same is true for other populations of chimps and gorillas. As anthropologists have been saying for years, human races are genetically meaningless, and the basis for racial differences is only a tiny part of our genome. In fact, evidence shows that most differences among humans, such as our skin color and the shape of our eyes, are the result of very recent changes in our evolution, occurring sometime after nearly all living non-African human lineages emerged from Africa, meaning only a few tens of thousands of years ago. This is an important thing to remember whenever issues of race come up.

So when did humans go through that bottleneck? The latest evidence gathered from dated archaeological sites places it at least 48,000 years ago, which is a bare minimum: there are not a lot of archaeological sites known from 48,000 or more years ago, so the bottleneck could have happened much earlier. But based on the age of dated archaeological sites and modern *Homo sapiens* fossils, we can use the molecular clock to get a minimum estimate for how long ago all the humans on the planet diverged from a small survivor population. Some recent genetic studies place this divergence around 70,000 years ago. Hmm: 70,000 years ago. What else was happening at that time?

5 Roots

The next time you visit a zoo, make a point of walking past the ape cages. Imagine that the apes had lost most of their hair, and imagine a cage nearby holding some unfortunate people who had no clothes and couldn't speak but were otherwise normal. Now try guessing how similar those apes are to us in their genes. For instance, would you guess that a chimpanzee shares 10 percent, 50 percent, or 99 percent of its genetic program with humans?

Jared Diamond, *The Third Chimpanzee*, 1992

The Third Chimpanzee

Before we talk in detail about what the Toba eruption may have done to humans about 74,000 years ago, we need to understand what kinds of humans had lived on the planet up until that time. Most people have only the vaguest notion of what human evolution is about, and much general understanding of the concept is full of misconceptions and false ideas that were outdated decades ago. Some people still think of human prehistory in terms of the grunting "cavemen" shown in many Hollywood movies and TV shows or the "modern Stone Age family" of *The Flintstones*. Despite these outdated stereotypes, anthropology has made huge strides in the past 30 years, uncovering thousands of human fossils and artifacts that together delineate a complex family tree that includes dozens of species and goes back 7 million years.

One of the most fundamental facts about humans is how closely related we are to the other great apes. As early as 1735, Carolus Linnaeus, the founder of modern scientific classification, placed humans among the other animals, giving us the scientific name *Homo sapiens* (thinking human). He diagnosed our species, in his *Systema Naturae*, with the Latin phrase *Nosce te ipsum*, "Know thyself." In 1766, George-Louis Leclerc, comte de Buffon, wrote in volume 14 of his *Histoire Naturelle* that an ape "is only an animal, but a very singular animal, which a man cannot view without returning to himself." Other French naturalists, such as Georges Cuvier and Étienne Geoffroy Saint-Hilaire, commented on the extreme anatomical similarity of apes and humans, although they refused to say that humans were a kind of ape. In 1809, the pioneering French biologist Jean-Baptiste Lamarck explicitly argued in his *Philosophie Zoologique* that

> certainly, if some race of apes, especially the most perfect among them, lost, by necessity of circumstances, or some other cause, the habit of climbing trees and grasping branches with the feet, . . . and if the individuals of that race, over generations, were forced to use their feet only for walking and ceased to use their hands as feet, doubtless . . . these apes would be transformed into two-handed beings and . . . their feet would no longer serve any purpose other than to walk.

Charles Darwin certainly appreciated the close connection between humans and apes, and it had a strong influence on his ideas. In 1838, a year after he returned from his landmark voyage around the world on the HMS *Beagle*, the young naturalist visited the London Zoo and saw its baby orangutan, named Jenny. He spent hours with her in her cage, noting her emotions and behaviors and how similar they were to those of humans in so many ways. In addition, he studied the zoo's other orangutans. In his unpublished notebooks, Darwin wrote extensively about their apparent tool use and creative play in fashioning toys out of sticks. He observed of Jenny that "she is fond of breaking sticks & [of] overturning things to do this (& she is quite strong) she tries the lever placing stick in hole & going to end as I saw.—She will take the whip & strike the giraffes, & take a stick & beat the men.—When a dog comes in she will take hold of anything, the keepers say, decidedly from knowing she will be able to hurt more with these than with paw."

Most impressive was the orangutans' self-awareness when they were given a mirror. As Darwin's notes describe two orangutans he observed:

> Both were astonished beyond measure at looking glass, looked at it every way, sideways, & with most steady surprise.—after some time stuck out lips, like kissing, to glass, & then the two did when they were first put together.—at last put hand behind glass at various distances, looked over it, <u>rubbed front</u> of glass, made faces at it—examined whole glass—put face quite close & pressed it—at last half refused to look at it—startled & seemed almost frightened, & evidently became cross because it could not understand puzzle.—Put body in all kinds of positions when approaching glass to examine it.

In 1839, when his first child, William, was born, Darwin did the same types of experiments, analyzing the child's emotions, expressions, and developmental landmarks and then comparing them to those of apes. He persisted in the tradition of scientifically observing his children when his first daughter, Anne Elizabeth, was born in 1841 but lost interest in the project with the eight additional children whom his wife, Emma, bore (three died in childhood). Darwin was deeply impressed by how many human emotions and behaviors the great apes showed, and thus he got past the superficial, sneering stereotype of apes as hooting, subhuman creatures that most people held at the time.

He also knew that most people were not ready to accept the idea that we are closely related to apes or, even worse, descend from an ape-like ancestor. He therefore tiptoed around the problem in his revolutionary 1859 book, *On the Origin of Species*, saying only that "in the distant future . . . light will be thrown on the origin of man." Of course, readers at the time were not fooled, and some decried his idea that "men came from monkeys," accusing him of blasphemy and heresy. He never publicly dealt with the issue directly until he published *The Descent of Man* in 1871. His main supporter and advocate, Thomas Henry Huxley, however, was not

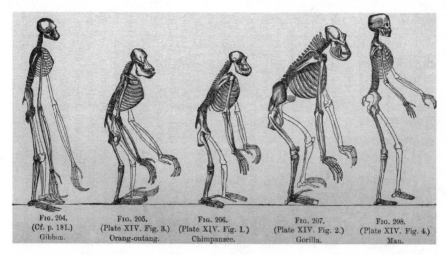

Figure 5.1. Thomas Henry Huxley's famous diagram of the bone-by-bone similarity between humans and great apes; our bones differ from theirs mostly in proportion. (From Huxley, *Evidence as to Man's Place in Nature*, 1863)

so timid. In 1863, Huxley published *Evidence as to Man's Place in Nature*, boldly showing the extreme similarities of the skeletons of humans and other great apes, which differ only in the proportions of their limbs and the shapes of their skulls (fig. 5.1).

In the decades since Huxley's book appeared, nearly everyone without religious blinders about humanity has learned to get past the image of screeching apes and see their humanlike features. Thanks to the work of Jane Goodall on chimps, Dian Fossey on gorillas, and Biruté Galdikas on orangutans, we have discovered just how humanlike these long-lost cousins really are. Chimpanzees and gorillas, as well as orangutans, can learn sign language, communicate in simple sentences, and make and use simple tools. Great-ape societies are very sophisticated compared with those of any other nonhuman animal and offer many insights into the complexities of human societies as well. More than a century of research by hundreds of anthropologists has documented numerous connections between apes and humans.

The similarities go far beyond anatomy and behavior. Neither Darwin nor any other biologist of his day could have known that there is another line of evidence that reveals our relationships to apes and other animals: DNA. Some of the very first molecular techniques demonstrated that our DNA and chimp and gorilla DNA are extremely alike. In 1967, for example, Vincent Sarich and Allan Wilson, at the University of California, Berkeley, put antibodies from humans and apes into the same solution and observed that they produced a far stronger immune reaction than was seen when the researchers combined human antibodies with those of any other animal. If our immune reactions are extremely similar to those of apes, then the genes that produced our immune system must also be very similar.

Also in the 1960s, biochemists developed a technique called DNA-DNA hybridization. Using a warm chemical mix, biologists heated DNA from humans and different kinds of apes until the molecule's strands "unzipped." After the mixture cooled, nearby strands bound together, producing hybrid DNA with one strand from a human and one from another animal, such as a chimp. When the hybrid DNA was heated again, researchers found that the more tightly bound together two strands are (reflecting their degree of genetic similarity), the higher the temperature that is required to unzip them. By hybridizing the DNA of chimps, gorillas, and other apes plus monkeys, lemurs, and other animals with that of humans, scientists can get a rough measure of how genetically similar these all are. To no one's surprise, DNA from both species of chimpanzee (the common chimp, *Pan troglodytes*, and the pygmy chimp or bonobo, *P. paniscus*) has proved to be virtually identical to ours.

More recently, PCR and other technological breakthroughs have made it possible to directly sequence the DNA of not only humans but also many other animals as well as plants. The human genome was first sequenced in 2000 and the chimp genome in 2005. When they were compared, researchers got the same result: humans and chimps share 98 to 99 percent of their DNA. Less than 2 percent of our DNA differentiates us from gorillas as well.

This incredibly close similarity is because about 60 to 80 percent of human, chimp, and gorilla DNA is junk that is never read or used but instead is carried around passively generation after generation (see chapter 4). Some of the junk DNA is from endogenous retroviruses, remnants of viral DNA that were inserted into the genome when some distant ancestor was infected and which are still carried around even though they no longer code for anything. A tiny percentage is genes, including ones no longer in use, that code for every protein and structure in the body. The 1 to 2 percent of our DNA that distinguishes us from chimps is known as regulatory genes. These are the "on-off switches" that tell the rest of the genome whether it should be expressed or silent. They explain why humans look so different from the rest of the apes despite our nearly identical genes. For example, all apes and humans carry structural genes for a long tail. Unlike our monkey relatives, however, apes and humans do not express those genes, except in rare cases when the on-off switches fail. Then it is indeed possible for a human to grow a long, bony tail.

Other examples are discovered all the time. Most people know that birds today have a beak with no teeth. Yet all the earliest birds, from the Age of Dinosaurs, had teeth shaped like simple cones or pegs, as did their dinosaur ancestors. Almost 40 years ago, scientists were surprised to find that living birds still have the genes to make teeth. They injected tissue from the mouth of a chick into developing mouse embryos, which then grew teeth: not the teeth of mice, though, but teeth like those of dinosaurs. The ancestral but unread genes in our DNA mean that a mistake in gene regulation can cause ancient features to reappear. Other experiments have unlocked the ancient genes for a long, bony tail in birds; the tailbones of living birds are fused into a pygostyle, part of the stumpy feature called the parson's or pope's nose. Yet another experiment produced chicks with feet like those of dinosaurs, not birds, and another produced birds with a dinosaurian snout rather than a beak. Birds carry nearly all their dinosaurian genes around with them, but most of the time they are not expressed.

As we've seen, human genes are extremely similar to those of chimpanzees, and this fact is some of the best proof of our intimate relationship. Despite the revulsion some people feel toward apes and monkeys, they are indeed our closest relatives. The biologist Jared Diamond proposed this thought experiment: Imagine that some aliens land on earth and collect DNA from many animal species, including humans, which they sequence to see who is related to whom. Based on DNA similarity alone, the alien biologists would conclude that humans are a third species of chimpanzee.

Our DNA and chimp DNA are more similar than the DNA of any two species of frog. Our genes are even more similar to chimp genes than those of lions and tigers are to each other. All it would take to modify a human into a chimp (or vice versa) would be tiny changes in the regulatory genes, which would yield huge results.

Case closed: humans are slightly modified apes. The evidence from our genes, as well as from our anatomy, is overwhelming. The DNA in every cell in our bodies is a testament and witness to our close relation to chimps, no matter that this fact might make some people uncomfortable. We would have known all this without having discovered a single fossil human that shows the transition from apes. But that raises an important question: how long ago *did* humans and apes diverge?

Fossils versus Molecules

The similarity of our genes is a function of how ago long the human lineage diverged from that of the apes. In the 1960s, paleoanthropologists working with fossils were convinced that the split had happened about 14 million years ago, based on a fossil then called *Ramapithecus*, found in the strata of the Siwalik Hills (in today's Pakistan) in 1932. Rama is one of the Hindu gods, and *pithecus* is from the Greek for "ape"; there are also primates named after the Hindu gods Shiva and Brahma. *Ramapithecus* had relatively small canines, and the shape of its jaw, when seen from the top, was more like a broad semicircle (a shape typical of human jaws) than the U of an ape jaw; it also had a flat chin. The Harvard anthropologist David Pilbeam, the Yale (later Duke) primatologist Elwyn Simons, and others championed the view that this jaw was the oldest known hominid fossil, meaning that it came from a member of our own family, Hominidae. Throughout the 1960s and 1970s, every student of primate evolution learned about *Ramapithecus* as "the first hominid." This was very much a part of the textbooks (and my classes) when I took anthropology and human paleontology in the 1970s.

Meanwhile, biologists were working hard on molecular clock estimates of the timing of divergence of many groups of animals (see chapter 4). Again and again, research by Sarich and Wilson showed that the divergence between humans and chimps happened only 5 to 7 million years ago and no more than 8 million years ago, not the 14 million years that anthropologists promoting *Ramapithecus* suggested. Paleontologists, too, were sure of the fossils, asserting that there must be a problem with the molecular clock method, because every once in a while it did give very strange and ridiculous results. This still happens today, and we don't always know why.

During the 1970s and 1980s, the debate grew intense, and both sides got into shouting matches at meetings and in journals. Sarich and Wilson were convinced that their data were reliable and that something was wrong with *Ramapithecus* or its age. I knew Sarich quite well; he was a burly, towering, impressive figure with a natty beard, a loud voice, and strong opinions. He didn't mind ruffling feathers and offending people if necessary. In 1971, in "A Molecular Approach to the Question of Human Origins," he wrote, with emphasis, "One no longer has the option of considering a fossil specimen older than about eight million years as a hominid *no matter what it looks like.*" This, of course, upset paleoanthropologists such as Simons and Pilbeam, who insisted that *Ramapithecus* showed that human divergence from apes happened much earlier than the molecular data suggested.

The solution to the conflict came from paleontologists who found more fossils. In 1982, Pilbeam himself reported on newly discovered specimens from the Siwalik Hills that included not only a more complete lower jaw of *Ramapithecus* but also a partial skull. With the addition of the skull, *Ramapithecus* looked much more like a fossil orangutan that Guy Pilgrim had named *Sivapithecus* in 1910. The lower jaw that had been named *Ramapithecus* was, in fact, just the jaw of a fossil relative of the orangutan that happened to look like a hominid. Eventually, anthropologists conceded their error, which gave the victory to Sarich, Wilson, and molecular biology.

The discovery that apes share tremendous amounts of anatomical and behavioral similarities with us, the revelation that our DNA is just a few percent different from that of our ape kinfolk, and the evidence that our lineage diverged from the apes only recently—all of these things have come together to place us with our close ape relatives in the animal kingdom, and to change how we think about ourselves as somehow "special" and separate from the rest of the animals. Scientific evidence shows that the difference is both slight and a matter only of degree—certainly not of kind.

Out of Eurasia?

When Darwin published *On the Origin of Species* in 1859, there was almost no human fossil record to support his case. The first Neanderthal remains had been discovered in 1856 and reported in 1857, found in a cave in the Neander Valley (*thal* in German) near Düsseldorf. However, they consisted only of a skullcap and a few limb bones, and they were originally mistaken for cave bear bones. Later, among other bizarre ideas, the fossils were widely dismissed as being the remains of a diseased Cossack cavalryman. In the early days, no one viewed the Neanderthal fossils as representing more than just an unusual modern human. In 1908, when the earliest complete Neanderthal skeleton was found, in La Chapelle-aux-Saints in France, it further confused the issue. It came from an old man with rickets, so the early reconstructions falsely pictured Neanderthals as the stooped and brutish "cavemen" of stereotype. However, better skeletons from healthy Neanderthals showed that they were fully upright and powerfully built, much stronger and heavier-boned than almost any modern human, with a flatter skull that sticks out in back and has strong brow ridges (fig. 5.2).

In Darwin's time, no fossil humans had yet been uncovered that looked very different from modern humans. The discovery of more-primitive hominid specimens that were clearly different from *Homo sapiens* would wait until the end of the 19th century. Nonetheless, in *The Descent of Man* (1871), Darwin suggests that humans must have evolved in Africa because all our ape relatives (chimpanzees and gorillas) came from there. Nearly every anthropologist for the next 50 years ignored this idea, however, and followed the belief that humans had first appeared in Eurasia.

There were a number of reasons for this, but paramount among them was a deeply held racism. In fact, the idea can be traced back to the

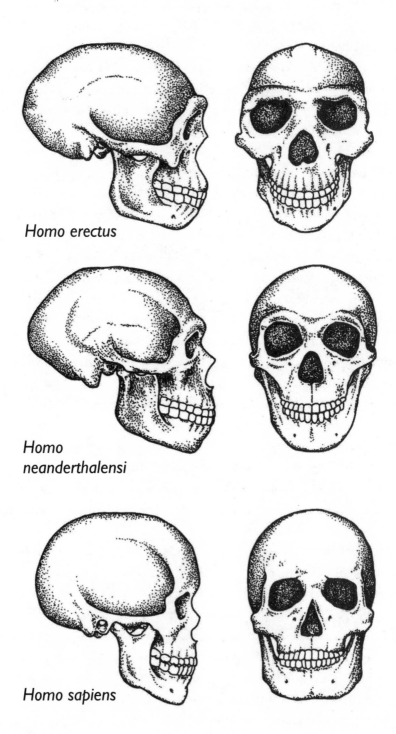

Homo erectus

Homo neanderthalensi

Homo sapiens

Figure 5.2. Comparison of skulls of *Homo erectus*, modern *Homo sapiens*, and a Neanderthal.

German embryologist and biologist Ernst Haeckel, who—even though he was Darwin's greatest protégé in Germany—forcefully argued for the Eurasian origin of humans long before any fossils had been found that might support his argument. Throughout the 19th and early 20th centuries, most anthropologists considered African peoples to be subhuman or apelike. Some did not consider Africans to be members of our species, *Homo sapiens*, at all. The idea that all humans descended from black Africans was abhorrent to the predominantly racist scholars of the time.

Eugène Dubois, a Dutch doctor and anatomist, was fascinated with Darwin's ideas, though he also subscribed to the notion that humans must have evolved in Eurasia. To obtain a chance to find fossils, he volunteered for the Dutch army as a surgeon and ended up in what was then the Dutch East Indies (now Indonesia). Most people would regard as foolish Dubois's choice of where to dig: he searched randomly in some river gravels without having any real clues about what fossils they might hold, but he was extraordinarily lucky. After a few excavations, between 1891 and 1895 he and his Javanese crews found a series of specimens: a skullcap, a thighbone, and a few teeth. Collectively, he called these *Pithecanthropus erectus* (from the Latin for "upright ape-man"), but they were better known as "Java man" because of where they were found. Although the specimens did not form a complete skeleton, it was clear from the thighbone that its owner had walked upright. The skullcap was very primitive, with prominent brow ridges, and showed that the brain volume was only half that of modern humans.

By 1896, after his professional isolation in far-off Java was over, Dubois had returned to Holland to resume his career as a professor of anatomy. Because the fossils he found were so incomplete, many anthropologists were not convinced of his claims. Some thought the remains were merely those of a deformed ape. Dubois, who had a prickly personality, was stung by the criticism and withdrew from the debate. He hid his specimens away and refused to show them to anyone or even to get involved in the scientific discussion about what they represented. Dubois remained withdrawn and embittered until his death in 1940 at age 82. However, opinion had begun to turn in his favor in the 1920s.

In that decade, the paleontologist Henry Fairfield Osborn, the president of the American Museum of Natural History in New York, sponsored the Central Asiatic Expeditions to Mongolia. There, he was sure, early human fossils would be found. The expedition didn't find any, but it was a huge success in discovering spectacular Asian dinosaurs such as *Velociraptor* and the first dinosaur eggs, as well as mammals such as the gigantic hornless rhinoceros *Paraceratherium*.

While scouting Asia for Osborn's expeditions, the American Museum of Natural History paleontologist Walter Granger saw apothecaries in open-air Chinese markets selling "dragon bones," which were ground into a powder for traditional Chinese medicine. He recognized that these bones were in fact mostly fossils of Ice Age mammals, including bears, hyenas, rhinoceroses, tapirs, cattle, and horses. The Swedish paleontologist Johann Gunnar Andersson also discovered this fact, and after snooping around he learned that the fossils came from a place called Dragon Bone Hill at the Choukoutien (Zhoukoudian) cave system, southwest of Beijing. When he and Granger located the caves, Andersson picked up some bone fragments and other promising clues, such as flakes of white quartz from artifacts, and told his companion, "Here is primitive man; now all we have to do is find him."

Andersson's assistant Otto Zdansky (an Austrian paleontologist) and a large Chinese crew started the first excavations, which produced not only the bones of gigantic hyenas but also fragments of other Ice Age mammals. This suggested that the nonhyena remains were remnants of meals that had been dragged into a hyena's den. Among them were two primitive hominin molars that Zdansky found in 1926. (Hominins are members of the tribe Hominini, which includes just the human lineage of our immediate ancestors. This is a group within the larger family Hominidae, which includes the human lineage and the great apes.) The Canadian anatomist Davidson Black, who was teaching anatomy at Peking Union Medical College, took over the supervision of the excavations and published a description of the teeth in 1927, announcing the discovery of *Sinanthropus pekingensis*, meaning "Chinese human from Peking." Even though his article appeared in *Nature*, many scientists were skeptical of a new species identification that was based on just two teeth.

Black then obtained money from the Rockefeller Foundation to greatly expand the excavations and hired many more Chinese workers to dig on a massive scale. The larger excavations produced a lower jaw, skull fragments, and more teeth, which confirmed the primitive nature of the new species. Eventually, the scientists had more than 200 hominid fossils, including six nearly complete skulls. The work continued until Black unexpectedly died at age 49 in 1935. The German anatomist Franz Weidenreich took over the study of the fossils. Although Black had published many preliminary descriptions of the fossils as they were found, it was Weidenreich's detailed monographs that provided complete documentation of the finds. With these new discoveries, it soon became apparent that "Peking man" was very similar to Dubois's Java man. Currently, most anthropologists consider them to be part of the same species, *Homo erectus.*

Meanwhile, the Japanese Empire was expanding throughout much of eastern Asia. In 1931, the Japanese had invaded Manchuria, in northeastern China near the excavations, and turned it into a Japanese province, Manchukuo. Puyi, the last Qing emperor of China, was installed as the head of a puppet government run by Japan (memorably depicted in the movie *The Last Emperor*). In 1937, Japan invaded again, annexing another large chunk of China even as it fought both the Nationalist army under Chiang Kai-shek and the Communists under Mao Zedong.

By mid-1941, the scientists at Zhoukoudian could see more signs of trouble on the horizon, as the Japanese had occupied Beijing. (The Pearl Harbor attack and America's entry into World War II would come just a few months later.) They worried that the precious fossils would become war souvenirs and be hauled off to Japan, never to be seen again. They packed all the specimens at Peking Union Medical College into two large crates, which a group of US Marines tried to smuggle out of the country through the port of Qinhuangdao. Ironically, in the midst of this panicked, secret rush to hide and protect the fossils from Japanese invaders, the crates were lost. No one knows where they went. Some say they were loaded onto an American ship that was later sunk by the Japanese. Others think they were captured and put on the Japanese ship *Awa Maru*, which the Americans later sunk. A few people think the fossils were secretly buried to prevent their discovery, but if so, their location has long since vanished, along with those who did the burying. Still others think Chinese merchants found and destroyed them as they did with other "dragon bones" used in traditional medicine. Fortunately, molds and casts had been made of nearly all the original material, and accurate replicas are found in many museums today, so we do know what these fossils looked like. In addition, more recent excavations at Zhoukoudian have found much more material, so the loss was not irreparable.

The Piltdown Hoax

The notion that early humans must have come from Eurasia, where the "superior" races had long lived, was so deeply ingrained in the early 20th century that a clever hoaxer could take advantage of it. In 1912, when the only known human fossils were Dubois's Java man and a few Neanderthals, an amateur collector named Charles Dawson announced that a worker had found a skullcap in a gravel pit near Piltdown, England, some four years earlier. The worker had thought it was a fossil coconut and planned to break it up, but Dawson saved the specimen. He went to the gravel pit himself and found more pieces, including a broken jaw.

The British Museum paleontologist Arthur Smith Woodward found nothing when Dawson brought him out to the pit, but Dawson gave him the specimens to piece together. The lower jaw was very apelike, but the skull looked like that of a modern human, with a large brain volume and small brow ridges. Crucially, the hinge of the jaw had been broken off, as were the face and many parts of the skull, so there was no way to tell if the jaw properly fit the skull. In August 1913, Dawson, Woodward, and the French priest and paleontologist Pierre Teilhard de Chardin returned to the Piltdown spoil piles, where Teilhard found a canine tooth that fit in the gap between the broken parts of the jaw. The canine was small and humanlike rather than being a large fang like the canine of an ape.

The Piltdown specimens did not receive universal acceptance. Sir Arthur Keith, a famous anatomist, thought the reconstruction was much too apelike and made one that looked more like a modern human. David Waterston, an anatomist at King's College London, argued that the specimens did not belong together and that "Piltdown man" was just an ape jaw found alongside a human skull. The French paleontologist Marcellin Boule, who had described the rickety Neanderthal skeleton found at La Chapelle-aux-Saints, also felt that these two specimens didn't belong together. Nor did the American zoologist Gerrit Smith Miller, who wrote in his "The Jaw of the Piltdown Man" (1915): "Deliberate malice could hardly have been more successful than the hazards of deposition in so breaking the fossils as to give free scope to individual judgment in fitting the parts together." Franz Weidenreich thought that it was a modern human skull incorrectly associated with an ape jaw whose teeth had been filed to make them look more human.

Nevertheless, prominent figures in British anthropology (Woodward, Keith, Sir Grafton Elliot Smith) supported the authenticity of Piltdown man as a real species (fig. 5.3). Their doubts were overcome by the biases of most anthropologists of that time. For one thing, the Piltdown skull suggested that the enlargement of the brain was what drove human evolution, long before hominins became bipedal or lost their apelike teeth and jaws. (We now know that upright posture and many of the dental and jaw features of hominins came before brain development.) These anthropologists all believed that the brain and intelligence were what defined humans and differentiated them from apes and other animals. Piltdown man was also even more primitive than Java man or Peking man, suggesting that Europe—and the British Isles in particular—was the center of human evolution. These British paleontologists were proud of the idea that the "missing link" or "the first Briton" came from their homeland.

Not until the early 1950s did most people begin to question the authenticity of the fossil. How did the hoax persist for so long? The main reason was that the British Museum kept the original specimens under lock and key and allowed scientists to study only replicas close up. Plus, Piltdown man fit all the prejudices claiming that advanced humans had

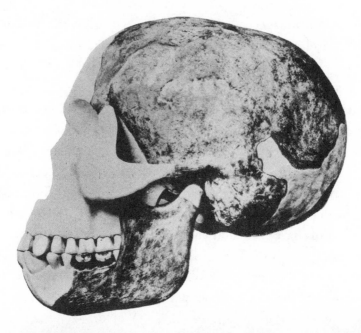

Figure 5.3. The Piltdown hoax. *a.* Replica of the specimens as restored, showing that only the skullcap and lower jaw were known and that key parts were missing. *b.* Leading British anthropologists and anatomists examine the specimen in this untitled 1915 painting by John Cook. Front row, left to right: A. S. Underwood, Sir Arthur Keith, W. P. Pycraft, and Sir Ray Lankester. Back row, left to right: F. O. Barlow, Sir Grafton Elliot Smith, the hoaxer Charles Dawson, and Arthur Smith Woodward. (Both courtesy Wikimedia Commons)

evolved in Europe, not Africa. By placing that evolution in Britain, it became the pride of British anthropology, so hardly anyone there dared to question it.

By the early 1950s, however, the Piltdown specimens had become an embarrassment: they didn't fit with the large number of fossils that had since been found and which showed that humans had evolved in Africa. Finally, in 1953, the chemist Kenneth Oakley, the anthropologist Sir Wilfred Le Gros Clark, and Joseph Weiner successfully insisted on seeing and testing the originals. Sure enough, the specimens were a hoax. The skull was that of a modern human, taken from a medieval grave, while the jaw came from an orangutan and some of the teeth from a chimp. The hoaxer had stained them all with iron and chromic acid to make them look both similar and old. As Weidenreich had suggested, the teeth had also been filed to disguise their orangutan source.

Since that time, there has been a veritable cottage industry of investigators trying to discover exactly who was behind the hoax. Dawson died in 1916, so he was gone before the skepticism over the fossils had a chance to implicate him. Naturally, as the person who "found" all the specimens, he is the main suspect. Investigation into his past showed that he had a long history of forging artifacts and human fossils from caves, so he could have been the sole culprit. Some people, however, think that he didn't have the anatomical skills to make a successful fraud or know enough to break off the jaw hinge region so no one could tell that the parts didn't belong together. Therefore someone with anatomical training must have helped him: Keith, Teilhard, Martin A. C. Hinton, Horace de Vere Cole, and even the author of the Sherlock Holmes stories, Sir Arthur Conan Doyle, have been suggested. We know that Dawson, with a major history of hoaxes and frauds, was clearly behind it, but whether he had help may never be known. It's been more than 100 years since he died, and the trail has gone cold.

Meanwhile, the search in Eurasia for humanity's earliest ancestors had proved to be a wild goose chase. It turned out that Darwin was right about Africa after all, as we will see in the next chapter.

6 Out of Africa

There is always something new out of Africa.

Pliny the Elder, *Natural History*, 79 CE

> We're all one dysfunctional family
> No matter where we nomads roam
> Rift Valley Drifters, drifting home genome by genome
> Take a look inside your genes, pardner, then ya
> Will see we've all got a birth certificate from Kenya.
>
> Roy Zimmerman, "Rift Valley Drifters," 2008

The Taung Child, Mrs. Ples, and Dear Boy

Given the discovery of Neanderthal specimens in the 1850s—and Java man in the 1890s, "Piltdown man" in 1912, and Peking man in 1927—the case that humans had originated in Eurasia seemed strong in the early 20th century. Of course, this was largely a consequence of the fact that most anthropologists were based in Europe and looked locally first—or, as in the cases of Dubois, Andersson, Zdansky, and Black, traveled to Java or China to find fossils. No one thought that there might be more primitive fossils in Africa, despite Darwin's suggestion that humans probably originated there because our nearest relatives among the apes are from Africa. Not only was Africa still a colonial backwater with few scientists and little scholarly research, but most of the continent didn't have universities or museums.

The exception was South Africa, which had been colonized and developed much earlier than most of the rest of Africa because of its enormous wealth in diamonds, gold, and many other valuable commodities. Both the Dutch settlers (Afrikaners) and the British colonists there had established large modern cities, including Cape Town, Johannesburg, Durban, and Pretoria. In fact, in 1931 South Africa was the first African country to gain its independence from its colonial masters, after becoming a dominion in 1909 (although it did not become a republic until 1961). Most of the rest of the African colonies did not gain independence until the 1950s or 1960s.

Among the top South African universities was the University of the Witwatersrand in Johannesburg, which had a medical school staffed by people trained at the best European medical schools. One was a young Australian anatomist, Raymond Dart, who had gotten his medical degree from University College London and then emigrated to South Africa in 1922. When he started at Witwatersrand, he found that its anatomy program had no human or ape skulls or skeletons, which are essential to teaching anatomy even today. He told his students that he was seeking interesting bones, hoping for a few modern and possibly fossil monkeys or baboons. Josephine Salmon, one of the few women in the class, told him that she had seen a fossil baboon skull on the mantel of a friend's fireplace. She brought it in, and indeed it was a baboon fossil. Her friend was the son of E. G. Izod, the director of the Northern Lime Company, which was quarrying limestone from a cave called Taung to crush into cement. Dart made the acquaintance of the elder Izod and asked him to send over any other fossils his crew might find at Taung.

One day in 1924, Dart was getting dressed, preparing to serve as the best man at a friend's wedding, when two huge crates were delivered to his house. The first held nothing remarkable, but when he pried open the second, he saw, right on top, the natural cast (endocast) of a fossil brain from something close to a human. It had formed when sediment had filled the empty brain cavity of a skull and then hardened into rock. Dart was thunderstruck by the specimen because the brain looked remarkably human, even though it was only slightly larger than that of a chimpanzee.

He hastily rummaged through the rest of the crate, bursting with excitement and delaying the start of his friend's nuptials, until he found the face that had once been attached to the brain. Then he went to the wedding. As soon as the festivities were over, he rushed back to his treasure. He spent the next few months carefully cleaning and preparing the skull, then extracting it from its hard limestone tomb. He even managed to split the rock coating off the face by making one delicate hammer stroke on the knitting needles that he used instead of a chisel.

Once the Taung specimen was clean, Dart wrote up a careful description, which he published in *Nature*. The fossil was the skull of a child with all its baby teeth still present, as well as a complete face and the complete brain endocast (fig. 6.1). Dart named it *Australopithecus africanus* ("African southern ape" in Latin) and argued strongly that it was a more primitive human fossil than any that had yet been found. He was particularly well suited to studying brain endocasts, since brains had been his specialty in medical school. He could tell that this one was much larger than that of an ape with a skull of the same size, and it had an advanced forebrain, more like a human's than like any of the apes'.

Dart expected his discovery to transform anthropology. Instead, most European scientists dismissed it as a "juvenile ape." This criticism was not entirely off base, because juvenile apes skulls have the proportions of a human adult (large brain, flat face, small canines, no brow ridges), so they do indeed look a lot more like humans than like adult apes. But Dart correctly pointed out that his specimen had numerous hominin features that could not be dismissed as those of a juvenile ape, including upright posture (which he determined from the position of the foramen magnum, the hole for the spinal column, below the center of the skull), small canine teeth, arrangement of those teeth in a semicircle or C pattern (rather than the U pattern of ape teeth), and an enlarged forebrain.

Nevertheless, the European anthropological community remained unconvinced. Scientists doubted the specimen not only because it was a juvenile but also because, as we saw in the previous chapter, the prevailing dogma was that brain enlargement came first in evolution and defined what it meant to be human. Upright posture, they believed, came

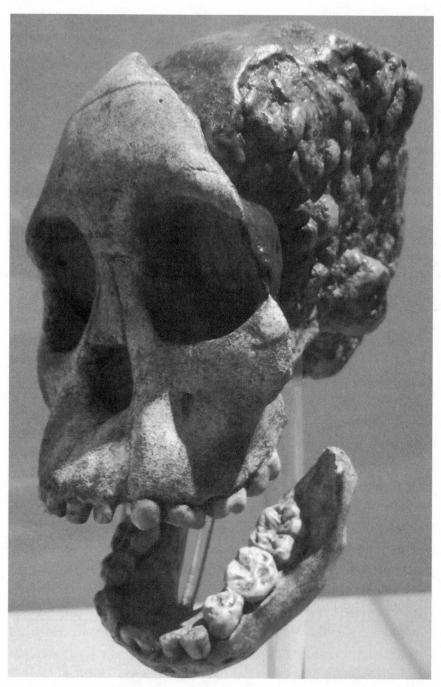

Figure 6.1. The skull and jaws of the "Taung child," showing the distinctive teeth, the flat face, the lack of brow ridges, and the endocast of the brain. (Photograph by Daderot of exhibit in Naturmuseum Senckenberg, Frankfurt am Main, Germany; courtesy Wikimedia Commons)

later. Indeed, the human fossils found so far (Peking man, Java man, Neanderthals) all had large brains, not that much smaller than those of modern humans. Even the Piltdown specimen, not yet debunked, with its apelike jaw and teeth, had a large brain. Yet the Taung specimen had a small brain but an upright posture, with a humanlike face and tooth row. In a 1925 letter to *Nature*, Sir Arthur Keith, one of Piltdown's biggest backers, wrote that Dart's "claim is preposterous, the skull is that of a young anthropoid ape . . . and showing so many points of affinity with the two living African anthropoids, the gorilla and chimpanzee, that there cannot be a moment's hesitation in placing the fossil form in this living group."

Other factors also encouraged resistance to accepting the Taung fossil as a "missing link." Dart was an obscure anatomist from a foreign medical school, not one of the top anthropologists in Europe, and they didn't know his reputation for competence. His claim by itself was shocking, and all the more so coming from a relative unknown. In addition, the original publication was very short: just four pages, with only three tiny photographs of the specimen. But even when Dart brought the specimen to Britain in 1931, it did not convince European paleoanthropologists. That was also the year when many of Davidson Black's illustrations and descriptions of Peking man reached Europe, overshadowing Dart and his seemingly outrageous claims. But the biggest barrier was the long-standing racial prejudice against Africa as the human homeland. In the minds of Europeans, it was the location of only the black races, whom they considered inferior to white-skinned Europeans.

What Dart needed was additional remains, especially of adult individuals, to overcome the objections based on the juvenile features of the Taung fossil. That problem was solved when the energetic Scottish doctor Robert Broom began a major collecting project in South Africa. He had already built his scientific reputation with his discoveries of 250-million-year-old reptiles and numerous fossils of the earliest relatives of mammals, which he found in the Karoo Desert of central South Africa. He had collectors and contacts all over the country, particularly in the limestone caves. In 1938, one of them, named Kromdraai, produced a very robust skull, with a much heavier, thicker construction than others uncovered to date. Broom called it *Paranthropus robustus* ("robust near-human" in Latin). Later, in the cave complex at Swartkrans, Broom and his crews found more than 130 *P. robustus* individuals. A recent analysis of their teeth has shown that none of these gorillalike humans lived past 17 years and that they ate a gritty diet of nuts, seeds, and grasses.

In 1947, Broom and the paleontologist John T. Robinson found the complete skull of what they thought was an adult female (although it is now believed to be a male) in the caves of Sterkfontein. It had gracile—lightly built and delicate—australopithecine features yet was just as primitive as the Taung child. This specimen, nicknamed Mrs. Ples (short for Broom's scientific name for the skull, *Pleisanthropus transvaalensis*, or "near-ape from the Transvaal"), became world famous. It showed that South Africa was producing much more primitive hominin fossils than anywhere in Eurasia. Soon additional fossils of many australopithecines were found in Sterkfontein, establishing the natural range of *P. transvaalensis*'s population variability. Some years later, anthropologists decided that the Sterkfontein adults and the Taung child are the same species, so *P. transvaalensis* is now lumped into Dart's original taxon, *Australopithecus africanus*.

Thus, by 1948 the most-primitive known hominin fossils had come from Africa, while nothing so archaic had been found in Eurasia. Darwin and Dart were right about humans originating in Africa. The discoveries there also proved that the notion that brains drove human evolution was dead wrong: brain enlargement was one of the last events in human prehistory, while bipedal posture, reduced canine teeth, and a shortened face were among the first things to appear in humans. This was why the Piltdown specimen became an embarrassment by the early 1950s. The revelation that it was a forgery was actually a relief to anthropologists. Piltdown was by then considered an anomaly that didn't fit into the emerging picture of human evolution.

Meanwhile, Dart had kept working on his fossils, and as the years went by, he gradually saw himself vindicated. In 1947, Keith admitted in a letter to *Nature*, "Dart was right and I was wrong." Dart lived until 1988, dying at the age of 95, celebrated and honored for his discoveries and for pioneering modern paleoanthropology. Most of his rivals had died long before and are now forgotten.

Next to Dart and Broom, the person who did the most to uncover our African roots was the famous and controversial Louis Seymour Bazlett Leakey. In my generation, the name Leakey was almost synonymous with "human evolution," thanks to coverage by *National Geographic* magazine and other media. The family tradition was carried on by his wife, Mary, his son Richard, and Richard's wife, Meave. By all accounts, Louis Leakey was a charismatic figure who inspired and mentored many generations of scientists, including three young women whom he convinced to change their career goals and devote their lives to the close examination of our ape relatives in the wild: Jane Goodall, who studied chimpanzees; Dian Fossey, who studied mountain gorillas (and was murdered, possibly by poachers); and Biruté Galdikas, who studied orangutans. Leakey was legendary for his determination and his ability to weave a spellbinding tale about his discoveries. He was also a scientific maverick who often challenged or scorned the anthropological consensus, and his critics regarded a lot of his work as somewhat careless and lacking rigor. More than once, he pushed controversial ideas that turned out to be wrong (such as the alleged Calico Early Man Site in California's Mojave Desert, which turned out to be not an archaeological site but a natural feature). Leakey was also famous for a streak of good fortune that allowed him (and his family) to make discoveries in places where no one had found anything for years. It was known as "Leakey's luck" in the trade, and it brought him success on a level that few other scientists have reached.

Even Leakey's background was unconventional. He was born in Kenya to British missionaries, so he held dual citizenship as a Briton and as a Kenyan. He grew up with African wildlife all around him and learned the rigors of life in the African bush better than any European could. He was adopted by the Kikuyu, one of the largest tribes in Kenya, and spoke their language and knew their customs well. After getting an education from tutors, he went to study at Cambridge University after the First World War. There he showed his native brilliance but also his stubbornness and eccentricities. He refused to do some things that his professors thought were important and declined to follow conventional ideas if he didn't agree with them. Nevertheless, he got his doctorate and had already published numerous papers about Kenyan archaeology when he pushed the establishment too far. In the 1930s, while still married to his first wife, Frida, he fell in love with the young artist Mary Nichol, who worked for him. Leakey abandoned his wife, and the scandal forced him to leave Cambridge and return to his native Kenya, where he finally got his divorce and then wedded Mary. Together they would find a number of sites that yielded fossils of primitive apes, in places such as Kanam, Kanjera, and Rusinga Island.

During World War II, Leakey used his fluency in Kikuyu and his connections with local tribes to spy for the British. He also acted as an interpreter and mediator between British colonial leaders and the Kikuyu (later he was instrumental in helping to mediate the pro-independence Mau Mau Uprising of 1952–60). When World War II ended, he refused to return to Europe, staying in Kenya and settling for a position with a tiny salary at the fledgling Coryndon Museum in Nairobi (now the Nairobi National Museum).

Louis and Mary Leakey discovered numerous fossil apes and did good archaeology but nothing that would catch the attention of anthropologists outside Kenya. He finally hit the mother lode in Tanzania's Olduvai Gorge. Originally studied by the German geologist Hans Reck in 1913, Olduvai had produced a fairly modern-looking human skeleton, much older than any found in Europe, that was found with early Ice Age mammal fossils. Scholars in Europe regarded it as a recently buried modern human, but in 1931 Leakey provided evidence, consisting of ancient mammal bones and stone tools, proving that it was indeed ancient. After 1951, he and Mary devoted their time to Olduvai, where they found numerous stone tools but no remarkable specimens.

Following eight years of hard labor at Olduvai (and 30 years after Louis first started working in East Africa), Mary found a remarkable skull in 1959. It was clearly far more primitive and robust than any fossil human yet discovered, with a large sagittal crest along its top and huge, crushing molars. The Leakeys nicknamed it Dear Boy. Others called it Nutcracker Man, for its big molars and robust jaw, but it was formally named *Zinjanthropus boisei*, or Zinj for short. Zinj was the medieval name of a region near the Tanzanian coast, and the species name honors Charles Boise, who funded the Leakeys' research. This skull was so remarkable that *National Geographic* featured it several times, and the Leakeys became world famous and never had to worry about money for their work again. Today the specimen is considered the most heavily built species of the robust australopithecine named *Paranthropus* by Broom, so formally it is called *Paranthropus boisei*.

More important, Zinj had an additional advantage: it could be accurately dated. The new method of potassium-argon dating was just spreading through geology in the late 1950s, yielding the first numerical dates for geologic events. Before then, geologists could merely speculate about how many millions or thousands of years ago things had happened. The discoveries from South African limestone caves could not be dated because those fossils were found as loose specimens amid undatable limestone debris; only a rough age could be estimated by means of associated mammal fossils. But the finds from Olduvai Gorge were different. The gorge cuts through numerous layers of not only fossil-bearing river deposits but also ash beds created by some of the many nearby volcanoes. Volcanic ash is the perfect substance to use for potassium-argon dating, so the Olduvai sedimentary sequence has been dated at multiple levels. Zinj came from Olduvai Bed I, the lowest level in the gorge, and was found immediately below a prominent volcanic ash bed.

Intrigued by the possibility of dating the oldest human fossils, the UC Berkeley geologists Jack Evernden and Garniss Curtis went to Kenya in 1960 and obtained an age for the ash layer immediately above Olduvai Bed I: more than 1.75 million years, far older than anyone considered possible. At that time, most scientists believed the entire Pleistocene was only a few hundred thousand years old. In the 1960s, a massive potassium-argon dating effort undertaken using rocks all over the world recalibrated the time scale.

Over the next few decades, more and more scientists swarmed to East Africa. They discovered numerous hominins, not only at Olduvai but also on the eastern and western shores of Lake Turkana (formerly Lake Rudolf), where Richard Leakey made his reputation. The Lake Turkana region also has a good layered sequence with volcanic ashes for dating. More recently, the best finds have been made in layered sediments with volcanic ashes in the Afar Rift region of Ethiopia along the middle Awash Valley. The well-dated Ethiopian fossils allow us to have great confidence in the time scale of human evolution in Africa.

The Human Lineage

Since the early work at Olduvai Gorge, the rate of discovery of new human fossils in Africa has accelerated ever more swiftly; typically only a few months go by before another amazing find is announced. Rather than relate each of these individual discoveries in the order in which they were found (which can be very confusing), I will discuss the current state of the human fossil record, from oldest to youngest, including its overall trends.

Currently, the oldest known fossil of our human lineage is a nearly complete skull from Chad found in 2001 by a team led by the French paleontologist Michel Brunet. Nicknamed Toumaï, which means "hope of life" in the Dazaga (Daza) language, the fossil is formally named *Sahelanthropus tchadensis*, after the Sahel, the sub-Saharan region where it was found, and the name of the country (spelled Tchad in French). The skull was crushed diagonally, so it is sheared and looks oddly asymmetrical in some views. Computer technology has allowed us to virtually undeform the specimen and make a 3-D replica for analysis.

The skull of *S. tchadensis* is about the same size as that of a chimpanzee, so it was assumed that this creature's body was the same size as a chimp's. However, its brain cavity is about 320 to 380 cubic centimeters in volume, larger than that of a chimp but smaller than the 1,350 cc of an average modern human. *S. tchadensis* has big brow ridges, like those found on chimps and primitive humans, and relatively primitive cheek teeth. Yet it also has many advanced human features: a flat face, without a snout, and reduced canines, resulting in both a tooth row and a palate shaped like a C. Most important, the foramen magnum is immediately below the center of the base of the skull, showing that *S. tchadensis* held its head upright and thus was bipedal and walked upright. Thus, bipedal posture, along with a flat face, reduced canines, and a C-shaped palate and tooth row, appeared at the beginning of human evolution. Contrary to the earlier expectations of anthropologists, these characteristics emerged long before the brain began to enlarge. Most important of all, the other mammal fossils found with the skull are from species that occurred 6 to 7 million years ago, in good agreement with molecular clock dates of the timing of the human-ape split (see chapter 5).

The next-youngest hominin fossils were found by a French-British-Kenyan team working in the Tugen Hills of Kenya and led by Martin Pickford. Volcanic ashes in the beds there date these fossils to between 5.7 and 6.1 million years old, just slightly younger than *Sahelanthropus*. The Tugen Hills specimens were first announced in 2000, and in 2007, additional fossils were described. They are called *Orrorin tugenensis*, after the hills. *Orrorin* is known from only about 20 specimens (mostly the back of the jaw, the front of the jaw, isolated teeth, fragments of the upper arm bone and the thighbone, and finger bones). The teeth, as far as they are known, are very apelike, but the hip region of the thighbone clearly shows that *O. tugenensis* walked on two legs.

Only slightly younger is the genus *Ardipithecus*. The oldest species is *A. kadabba*, discovered in 2001 in deposits in Ethiopia dated to between 5.54 and 5.77 million years ago. It consists mostly of fragments of jaws, teeth, and some parts of the skeleton. Much better known is *A. ramidus*, from Ethiopian beds dated at 4.35 to 4.45 million years old. Originally described from fragmentary fossils in 1995, it was relatively poorly known until 2009, when Tim White and his colleagues announced that they had found specimens from at least nine individuals, including one they nick-named Ardi, which has a nearly complete skeleton and skull. This makes Ardi the earliest hominin known from nearly complete material. Based on these fossils, *A. ramidus* was about the size of a bonobo or pygmy chimp, with a brain volume of only 300 to 350 cc. Like the other early hominins, it had small canines and a C-shaped palate and tooth row. Unlike later human relatives, it had a large big toe, adapted to grasping branches with its feet, suggesting that it spent most of its time in trees rather than on the ground. Its molar teeth are unspecialized, low crowned, and simple, very similar to those of apes, suggesting that it ate an omnivorous diet with a lot of fruit. The larger sample of fossils, including both presumed males and presumed females, shows that there was no big difference between them in size or canine teeth, unlike in chimps and nearly all other living apes. To some anthropologists, this suggests that very little male-versus-male competition or aggression, both traits typical of the other living apes, occurred among *A. ramidus*.

In Kenyan beds between 3.9 and 5.25 million years old, anthropologists have found the oldest known fossils of Dart's genus *Australopithecus*. The oldest species is *A. anamensis*, known first from a fragment of an upper arm bone found in 1965 and then from a jaw found in East Turkana in 1995 by Meave Leakey and her crew, who named the species that year. Since then, many more specimens (including a thighbone) have been uncovered in Ethiopia. Currently, there are fragments of both upper and lower jaws, part of a skull, and a shinbone. Based on these limited specimens, *A. anamensis* was about the size of a chimp, but the wear on its teeth suggests it was more of a leaf eater, like a gorilla. Indeed, the available parts of its limbs suggest that it was primarily bipedal but was also a tree climber like Ardi. Most of the specimens in Kenya and Ethiopia were found in beds that were deposited in a wet, forested habitat, not an open savanna.

The best-known *Australopithecus* is the partial *A. afarensis* skeleton nicknamed Lucy. *A. afarensis* fossils have been found in beds in Ethiopia dated between 2.9 and 3.9 million years old, and other finds that may also be *A. afarensis* have been uncovered in Tanzania. The species became famous with the publication of Donald Johanson and Maitland Edey's 1981 book *Lucy: The Beginnings of Humankind*. In 1973, the rival Leakey group had blocked Johanson and the budding young scientist Tim White (the UC Berkeley anthropologist who later found *Ardipithecus* and many other fossils) from collecting in Kenya. So the two Americans instead arranged to work with the French geologists Maurice Taieb and

Yves Coppens and the American anthropologist Jon Kalb in Ethiopia's Afar Triangle. They found only hominin fragments that year, their first field season.

On November 24, 1974, during their second field season, after spending months exploring and prospecting for fossils, Johanson took a break from writing field notes to help his student Tom Gray search an outcrop. Johanson spotted a glint of bone out of the corner of his eye, dug out the fossil, and immediately recognized it as a hominin bone. He and Gray dug out more bones until they had found almost 40 percent of a complete skeleton. It was the first partial skeleton of any hominin older than late Pleistocene Neanderthals. That night, as the company celebrated over a campfire, they listened to a Beatles tape. When "Lucy in the Sky with Diamonds" began to play, crew member Pamela Alderman sang along lustily and suggested that they nickname the fossil Lucy.

The next year, Johanson, White, and their crew went back to Hadar, on the southern edge of the Afar Triangle, where they found a big assemblage of *A. afarensis* bones preserved in an ancient mudflow deposit. Nicknamed the First Family, it was the first large sample of juvenile and adult hominin specimens from beds this old and gave anthropologists a look at how much variability was typical of a single population—important in deciding how much variation can be assigned to a single species. Based on this sample and Lucy's nearly complete skeleton, we know that *A. afarensis* walked fully upright (because of the shape of its hips and knee joint) and was more of a ground dweller than arboreal (its foot has no grasping big toe), though its shoulder blade, arms, and hands show that it was still capable of a lot of tree climbing. Its brain capacity was only 380 to 430 cc, and it had small canines yet an apelike snout.

Several other fossil hominins are known in the genus *Australopithecus*: *A. bahrelghazali*, from beds in Ethiopia around 3.5 million years old; *A. deyiremeda*, from beds in Ethiopia about 3.3 million years old; *A. garhi*, from beds in Ethiopia around 2.5 million years old; and *A. sediba*, from beds in South Africa about 1.78 million years old. *Australopithecus*, first reported by Dart in 1925 based solely on the Taung child, is now one of the most species-rich genera of all the hominins. With at least seven identified species spanning from 1.78 million to 5 million years ago, it is the longest-lived hominin genus as well. Some say it is too broadly defined and has become a sort of taxonomic wastebasket for all the primitive gracile hominins that are more advanced than *Ardipithecus* but not members of the genus *Homo*, but that is an argument for specialists.

In addition to all these members of *Australopithecus*, we should take note of the three species of robust hominins that used to be assigned to that genus (some anthropologists still place them there). Most experts on human fossils agree that they should be clustered in their own genus, *Paranthropus*. The first is *P. robustus*, whose original specimens were found at Swartkrans, South Africa, by Broom, described in 1938, and dated at around 2 million years old. The second is *Zinjanthropus* (now *P. boisei*), based on the hyper-robust skull found by the Leakeys at the bottom of Olduvai Gorge and dated at 1.8 million years old. The third is represented by the much older and much more mysterious "Black Skull." Dated at 2.5 million years old and described by my friend Alan Walker (since deceased) in 1985, it has large brow ridges, robust, flaring cheekbones, a strangely flattened face that protrudes from the skull, and the largest sagittal crest ever found in a hominin. The teeth are missing, but the size of their sockets shows that the molars were huge and grinding, like those found in other robust hominins. Today this specimen is known as *Paranthropus aethiopicus*.

The big sagittal crest, flaring cheekbones, and other features of the robust hominins named *Paranthropus* are required for the attachment of large jaw muscles and consistent with a lifestyle involving heavy chewing, perhaps of a fibrous, leafy diet like that of today's gorillas or one that included more seeds and nuts. Many of the specimens have extremely worn teeth, indicating a very gritty, tough diet. In fact, the big differences in the molars is one of the key lines of evidence that leads most paleoanthropologists to assign the robust species to their own genus, *Paranthropus*, rather than lumping them into *Australopithecus*.

Traditionally, it was thought that there could be only one species of human on the planet at a time, because there is only one human species alive now: *Homo sapiens*. As a result, for decades scientists tried to shoehorn all the variability of African hominins into just a few species. They dismissed the differences between gracile and robust forms as being similar to the differences seen between the robust skulls of male gorillas and the more gracile skulls of female gorillas. But when it became clear that there were at least seven types of gracile forms and three types of robust forms of African hominins, all with different ranges in time and space, this idea could no longer stand. In fact, in some levels in Lake Turkana, fossils from four different species have been found in the same beds. In some population samples (such as the one of *A. africanus* that includes Mrs. Ples), it is possible to distinguish male and female forms within a gracile species—so robust forms are not males of gracile forms but instead are of a different species. What is really striking about the robust forms is their molar teeth, which are much larger and more adapted to grinding than those found in the gracile australopithecines. Sexual differences can sometimes include body size, but rarely do males and females of the same species show big differences in the size and shape of their molar teeth.

The abundance of species of *Ardipithecus, Australopithecus, Paranthropus*, and so on found in the same beds shows that human evolution has not been a steady, linear march of progress in which one species follows another. Instead, it has formed a branching, bushy pattern with many lineages diverging multiple times and different species of the same lineage coexisting (fig. 6.2). This is demonstrated in eastern and southern Africa, where, in the interval between 1.8 million and 2.0 million years ago, *A. africanus, A. sediba, P. robustus, P. boisei*, and two species of *Homo* appear to have overlapped in time. *P. robustus* and *A. africanus* have been found in caves of similar ages in South Africa, while *P. boisei* and early *Homo* co-occur in the lowest beds at Olduvai. And, as if this huge diversity of overlapping species were not enough, there is yet another genus of hominin, *Kenyanthropus*, which was found by Meave Leakey's crew in beds about 3.5 million years old on the west shore of Lake Turkana. Its shape is different from that of any contemporary *Australopithecus*, especially its unusually flat face. Exactly how it fits with all the other species of hominins from that period is still a topic of debate.

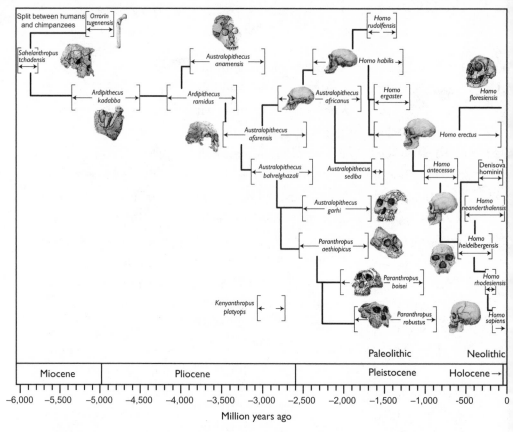

Figure 6.2. Family tree of hominin species, showing the branching, bushy history of human evolution. (Redrawn from several sources)

Our Genus, *Homo*

In addition to multiple branches of the *Australopithecus*, *Paranthropus*, and *Kenyanthropus* lineages, there are fossils that most anthropologists agree are members of our own genus, *Homo*. The criteria for distinguishing our genus from more primitive hominins are still not agreed upon, but they usually focus on the expansion of the brain. The earliest *Homo* specimens usually had a cranial capacity of at least 600 cc, compared to about 450 cc in the australopithecines. They also tended to be even more gracile than most australopithecines, with smaller brow ridges, no sagittal crest, and a bulging cranium, a consequence of their large brain size. In addition, the oldest specimens are associated with the primitive stone tools at Olduvai Gorge (products of what has been called the Oldowan culture), and it has long been thought that these humans were the first to make and use stone tools.

The oldest known *Homo* fossils come from several localities that have also produced fossils of *Australopithecus* and other primitive hominins, so the *Homo* branch overlapped in time with many of its ancestors. Originally, the oldest *Homo* species was based on fossils known as OH 7 (an abbreviation of Olduvai Hominid 7, its full name in the Leakey catalogue). They came from the bottom of Olduvai Gorge, the same level that produced *Zinjanthropus*, in 1960 and were dated at 1.75 million years old. The specimen was nicknamed Johnny's Child, as Jonathan Leakey, Louis and Mary's eldest son, found it while working with her.

Its fragments consist of a lower jaw, part of the skull roof, and 22 hand, finger, and wrist bones. The specimen was described and named *Homo habilis*, which translates to "handy man," by Louis Leakey, Phillip Tobias, and John Napier in 1964. Its cranial capacity was around 600 to 710 cc, and it had small cheek teeth compared to those of *Paranthropus* or *Australopithecus*, yet it also had curved, long fingers like those of a chimpanzee, suggesting that it was mostly arboreal. The initial discovery was controversial because the material was so limited, but nearly complete skulls were found later. The first, discovered in 1968 by the Leakey team and dated at 1.8 million years old, is known as OH 24, or Twiggy, and had a brain volume of 600 cc. In 1973, another one was found at Koobi Fora, on the eastern shore of what was then Lake Rudolf, and dated at 1.9 million years old. It is known as KNM ER 1813 (for the museum catalogue abbreviation of Kenya National Museum–East Rudolf 1813) and is relatively complete and undeformed, although its cranial capacity was only about 510 cc.

Most famous of all the *Homo* fossils is the nearly complete skull known as KNM ER 1470, dated to 1.9 million years ago. Richard Leakey's team found it at Koobi Fora in 1972, and it established his reputation in paleoanthropology independent of his famous father. Its relatively large, bulging braincase (with a cranial capacity of 700 cc), flat face, and smaller brow ridges make it look much more modern than any other fossil of similar age. In more recent years, some scientists decided that KNM ER 1470 is a different species than *Homo habilis*, and they named it *Homo rudolfensis*. Fossils found by Meave Leakey and her crew in 2012 seem to support the idea that *H. rudolfensis* is distinct from *H. habilis*. Many other scientists, however, including Tim White, do not regard *H. rudolfensis* as being any more than a variation of *H. habilis*. Meanwhile, in 1991, a jawbone was found in Malawi, dated to 2.3 to 2.5 million years ago, and identified as *H. rudolfensis*. In 2015, a lower jaw found in Ethiopia was assigned to *Homo*, and it dates to 2.8 million years ago; if its assignment to *Homo* is indeed correct, this is the oldest fossil of our genus yet discovered.

Modern Humans Evolve

By the late 20th century, most paleoanthropologists had come to agree that both Java man, discovered by Eugène Dubois the 1880s, and Peking man, found in China in the 1920s (see chapter 5), were the same species and that the species was a member of our genus. Thus Dubois's *Pithecanthropus erectus* became *Homo erectus*, as did *Sinanthropus pekingensis*. This hominin was bipedal and stood erect, as its species name implies, although later discoveries have shown that hominins have stood at least somewhat erect and been at least partially bipedal since the time of *Sahelanthropus*, almost 7 million years ago. *H. erectus* had about the same size range as modern *Homo sapiens* as well. Certain individuals, such as the nearly complete skeleton from West Turkana, Kenya, known as the Nariokotome Boy or the Turkana Boy, reached 190 centimeters (6 ft) in height. (Some anthropologists consider this and a few other specimens to be members of a separate species, *Homo ergaster*.) *H. erectus* had a brain capacity of about 1,000 cc, only slightly less than our own. Like earlier members of *Homo*, this species made crude choppers and hand axes; theirs are called Acheulean culture tools. Evidence also shows that they knew how to make and use fire and how to construct stone and wooden dwellings.

H. erectus appeared in Africa, where all of our other ancestors had long lived, about 1.9 million years ago. Anthropologists have evidence that by around 1.8 million years ago, the species had dispersed from the African homeland, for there are specimens from Indonesia (such as Java man) dated to that time, and fossils found elsewhere in Eurasia, such as Romania, are almost as old. There are abundant fossils of *H. erectus* that are about 500,000 years old from many parts of Eurasia, including Peking man, which has been dated as old as 460,000 years. The dating of the youngest fossils suggests that *H. erectus* may have persisted as recently as 70,000 years ago, overlapping with the Neanderthals and even some modern *H. sapiens*. *H. erectus* was thus the first member of our family to live outside Africa, and it roamed throughout the entire Old World (except the glaciated regions). It was also one of the most successful and long-lived species, existing for more than 1.8 million years. During most of that time, it was the only species of *Homo* on the planet, and its brain size and body proportions changed very little. Given this longevity, it could be argued that *H. erectus* was even more successful than we have been.

From that point on in human evolution, the picture is a bit cloudier. Many skulls and skeletons that begin to resemble modern humans have been found in localities all over Eurasia and Africa. When just a few fossils of our genus were known, researchers lumped them all under one general term: "archaic *Homo sapiens*." However, the large number of specimens discovered in recent years has allowed paleoanthropologists to recognize distinctive species groups among them, with distinctive cultural characteristics as well. A number of anthropologists have proposed new schemes for identifying various slightly different groups of specimens as distinct species; names have been restored to some species that were first given to them decades ago, before they were lumped into *H. sapiens*, and some proposed names are brand-new.

The first wave of *H. sapiens* to migrate out of Africa—sometimes identified as comprising distinctive species, such as *Homo antecessor* or *Homo mauritanicus*—spread across Eurasia from 1.2 million to 800,000 years ago, replacing *H. erectus*, which was already there in many places. Archaic *H. sapiens* specimens found in Eurasia and dated a bit later, to about 700,000 years ago, were given a distinct species name, *Homo heidelbergensis*. This species was originally based on a single jaw, called the Mauer mandible, that was found near Heidelberg, Germany, in 1907, but now it is represented by numerous skulls and skeletons found all over Eurasia, especially in the currently active excavations at the Sima de los Huesos (Pit of Bones) cave in the Atapuerca Mountains of northern Spain.

H. heidelbergensis may have originated from the African taxon *Homo rhodesiensis*, based on the evidence of the Kabwe 1 skull, known as Rhodesian Man, which was found in 1921 in the Broken Hill mine in Zambia (then called Northern Rhodesia). This skull was later matched with numerous others found at sites all over Africa. These specimens date to between 125,000 and 300,000 years old. Many anthropologists consider the *H. rhodesiensis* and *H. heidelbergensis* group to be, collectively, the ancestor of all the later human groups that arose over the past 400,000 years: the Neanderthals, the Denisovans, and possibly the "hobbits" of the island of Flores in Indonesia (discussed later in this chapter).

By about 400,000 years ago, another species had established itself in western Europe and the Near East: the Neanderthals. As chapter 5 describes, the first human fossils to be discovered were Neanderthals, although scientists initially dismissed them as merely the remains of a diseased modern Cossack who had died in a cave. Then, as luck would have it, the first complete description of their skeleton was based on an individual who had suffered from old age and rickets, so for decades Neanderthals were thought to be stoop-shouldered and primitive, the classic grunting "cavemen." However, modern research on healthy Neanderthal skulls and skeletons has shown that they were in truth very different from that stereotype. Neanderthal skulls are easily distinguished from ours: they had a protruding face, large brow ridges, no chin, and an occipital bun, a conical point on the back of the braincase. In a big blow to our ego, Neanderthals had a slightly larger average brain capacity than we do, and their cultural practice included ceremonial burials, suggesting complex religious beliefs. Evidence supporting this idea has been found in Shanidar Cave in Iraq, where Neanderthal gravesites have been discovered that were once surrounded by rings of flowers. To some researchers, these burials suggest that these Neanderthals possibly believed in an afterlife. They also cared for injured and otherwise incapacitated members of their group.

Neanderthals were robust, muscular, and slightly shorter than the average modern human. They lived exclusively in cold climates—the glacial margin of Europe and the Middle East—where their stocky build, rather like that of a modern Inuit or Laplander, would have been an advantage in that it helped them to retain body heat. Like their cultural practices, their tool kit, known as the Mousterian culture (similar tools were also used by *Homo sapiens*) was more intricate than those of previous *Homo* species, including small hand axes, spearheads, and arrowheads as well as bone and wooden tools. Some of their tools show complex workmanship and simple carving, so they were artistic in a manner unknown in earlier hominids.

For decades, anthropologists treated Neanderthals as a subspecies of *H. sapiens*, but work in the past 20 years suggests that they were a distinct species. The original evidence came from the Es-Skhul and Qafzeh Caves in Mount Carmel, Israel, where layers bearing Neanderthal remains alternate with layers containing the remains of early modern humans. In addition, Neanderthals appeared later than the earliest archaic *H. sapiens*, so they could not be our direct ancestors but rather must be an extinct European side branch. In 2010, their DNA was sequenced, clearly revealing that they were not *H. sapiens*. Surprisingly, genetic evidence also shows that all modern humans who are not of African descent have a bit of Neanderthal DNA, so there must have been some interbreeding between the two species.

Mysteriously, Neanderthals vanished about 40,000 years ago. Whether it was *H. sapiens* or some other cause that wiped them out is a highly controversial question. Pat Shipman argued that modern *H. sapiens* were the first to domesticate dogs, which gave them a big advantage over the Neanderthals in hunting and warfare. Whatever happened, modern *H. sapiens* soon took over the entire Old World, developing the complex Upper Paleolithic cultures collectively known as Cro-Magnon based on their art and artifacts. These modern *H. sapiens*, who were not a distinct species from the humans of today, created the famous cave paintings of Europe and fashioned many kinds of weapons and tools.

Other than living *H. sapiens*, Neanderthals were the last surviving species of *Homo* known to anthropology until 2010, when molecular biology shocked the world with the announcement that there was yet another species of human around in the past 41,000 years. Excavating Denisova

Cave in the Altai Mountains of Siberia, near the Russian-Mongolian-Chinese border, Russian archaeologists found a juvenile finger bone, a toe bone, and a few isolated teeth from a hominin, mixed with artifacts including a bracelet. The artifacts have a radiocarbon date of 41,000 years, so the age is well established. But when the molecular biology lab of Svante Pääbo and Johannes Krause at the Max Planck Institute for Evolutionary Anthropology in Leipzig, Germany, analyzed the mitochondrial DNA of the finger bone, they found it had a unique genetic sequence, distinct from those of both Neanderthals and modern humans. The nuclear DNA of the Denisovan specimens was also distinct but suggested that these people were closely related to the Neanderthals. They may have interacted with modern humans, because they share about 3 to 5 percent of their DNA with today's Melanesians and Australian Aborigines. Their mitochondrial DNA suggests that they branched off from the human lineage about 600,000 years ago and represent an "Out of Africa" migration that is separate from both the much earlier *H. erectus* exodus of 1.8 million years ago and the far more recent *H. rhodesiensis–H. heidelbergensis* emigration of 300,000 years ago.

These mysterious people are now referred to as the Denisovans. Since so few fossils of these people have been found, we cannot say much about their physical appearance or any other characteristic, except that they had distinctive DNA whose complete sequence is found in no other human species. In fact, most scientists are reluctant to give the Denisovans a formal scientific name, because there is not enough fossil material to describe the species in any normal sense. So this people remains obscure, showing us not only that the shape of the bones does not tell the whole tale but also that there may have been numerous other human species on this planet that didn't leave a fossil record.

Almost as shocking as the 2010 discovery of the Denisovans was the 2003 announcement of a dwarfed species of human found only on the island of Flores in Indonesia. Uncovered at the Liang Bua cave, their fossils and artifacts are dated to between 50,000 and 190,000 years ago. The most striking feature of these people is their tiny size: a full-grown adult was only about 1.1 meters (3 ft, 7 in) tall, which has won them the nickname "hobbits." Yet this species was not a tiny but fully modern human, like modern African Pygmy people, who have large brains proportional to their body size. Instead it was an entire population of small-brained hominins that appears to have descended from an *H. erectus* (or possibly even *H. habilis*) ancestry about a million years ago and which became dwarfed after it had arrived on the island. Size reduction is a common phenomenon on islands: many groups of animals, especially mammoths and hippos, have undergone dwarfing many times in their evolution, on islands ranging from Malta to Crete to Cyprus to Madagascar to the Channel Islands near Santa Barbara, California. The reason is clear: when the animals lived on the small resource base of an island, not the wider one of the mainland, they could not get the kind of nutrition needed to grow to formerly normal sizes. At the same time, their island setting spared them the pressure of large predators. Although interpretation of the "hobbit" fossils is controversial to a few scientists, most anthropologists agree that they do represent a distinct human species, which has been formally named *Homo floresiensis*.

The story of human evolution is complicated, with many species appearing over time, but the general picture of its branching, bushy tree is well established. Thus we know that when Toba erupted about 74,000 years ago, most of the species mentioned in this chapter were long gone. The handful still around included the Neanderthals, on the fringes of the glaciers in Europe; a variety of populations of modern humans in Africa and Eurasia; the "hobbits" on Flores; and the last of *H. erectus* in Eurasia. Interestingly, the last two species in this list seem to have vanished shortly afterward. The Toba eruption's effects on mainland populations of modern *H. sapiens* cannot be seen in the handful of skeletal fossils and scarce artifacts that we have discovered from around that time—but they *can* be detected in our DNA, and that is the subject of the next chapter.

7 Humanity at the Crossroads

I had a dream, which was not all a dream.
The bright sun was extinguish'd, and the stars
Did wander darkling in the eternal space,
Rayless, and pathless, and the icy earth
Swung blind and blackening in the moonless air;
Morn came and went—and came, and brought no day . . .
All earth was but one thought—and that was death
Immediate and inglorious; and the pang
Of famine fed upon all entrails—men
Died, and their bones were tombless as their flesh . . .

Lord Byron, "Darkness," 1816

Consilience

One of the most powerful moments in science is when a number of lines of investigation come together to produce a startling result. For example, in 1859, Charles Darwin showed that all available threads of evidence—breeding studies of domesticated animals, the hierarchical nature of classification, embryology, biogeography, vestigial organs, the fossil record, and much more—pointed to the inescapable conclusion that life evolves. The geologist Charles Lyell spent years collecting field evidence from around the world that converged to show that the earth is immensely old and that it has changed very slowly. In the twentieth century, once the method of using radioactive decay to measure the age of rocks was invented, scientists kept discovering older and older rocks until they reached a general agreement that the solar system's age is about 4.567 billion years. This date was based on two sources, moon rocks and meteorites, which have yielded the very oldest dates, and no measurement we have made since those sources' original analyses has exceeded that age limit. In cosmology, various threads of evidence—including Edwin Hubble and Milton Humason's 1920s discovery of the fact that the galaxies are retreating and that the universe, correspondingly, is expanding—pointed toward some sort of originary "big bang." Then, in the late 1950s, Arno Penzias and Robert Wilson accidentally found the predicted cosmic background radiation that independently confirmed that the Big Bang happened (see chapter 1). Many other great discoveries were made in the same way: with many independent lines of evidence converging to suggest a single conclusion.

The basic idea of a convergence or concordance of evidence goes back to the ancient Greeks. However, it was the British philosopher William Whewell who coined a specific term for this phenomenon in 1837: *consilience*. It comes from the Latin roots *con-* (together) and *siliens* (jumping), suggesting a moment when varying lines of evidence jump together and agree. (The word *resilience*, literally "jumping back," also has the latter root.) Brilliant in many areas, from mathematics to philosophy to poetry to architecture, Whewell spent his entire career at Trinity College, Cambridge, where two courtyards are named after him. He not only gained fame for his ideas about the philosophy of science but also coined many new words, including *scientist*, in 1833, to replace the old term *natural philosopher*.

Whewell pointed out that when multiple lines of research converge on a common answer, this is powerful evidence that researchers are on the right track and that a startling new perspective may be emerging. Consilience involves an observation of phenomena, and it has occurred in many areas of scholarship beyond the cases listed above, including historical investigations. For example, if a researcher finds several ancient texts all saying that a certain king began his rule of a particular country in

100, then the conclusion that he did establish rule in that place in that year is well supported. If the sources don't agree on the time or the place, the conclusion is suspect. Or take neo-Nazis who deny that the Holocaust happened or claim that it killed far fewer people than the 6 million Jews, Poles, Gypsies, and others whom it actually sent to their deaths. They try to support their case by nitpicking the number of people who were sent to concentration camps or taking selected quotes out of context. But the case against their deceit is solid because it is based on many different kinds of evidence: we have not only the memories of hundreds of survivors (now sadly fewer as they die in old age) but also the actual camps, their artifacts, and even the Nazis' documents themselves, which brag about their efficiency in killing.

As a practicing scientist, I have done my fair share of finding consilience, especially regarding the link between climate changes in North America 33 million years ago and the concurrent global cooling and birth of the Antarctic glacial ice cap. When I first verified that the changes in the earth's magnetic field can help to date rocks that yield important fossils, the thrill of discovery was intense. My crew and I would spend tedious months in the field collecting samples for analysis, which I'd trim over several weeks on a rock saw or grinding wheel into a shape that a machine known as a magnetometer could automatically load. This device measures the direction and strength of the planet's magnetic field as recorded by any rock at the time when it formed. As the samples slowly ran through the machine, my student collaborators and I would look at the display monitor, and even just the first data would tell us whether we had good results, and a likely publication. If we saw certain numbers, we knew that we had discovered a new magnetically reversed interval in a key rock sequence, which would help us to date those rocks and any associated fossils and place them in the context of the climatic events we were studying. We had unlocked the secrets in those rocks and gotten the first-ever reliable dates on them. When something like this happens in the life of a scientist, it's a thrilling "Ah-hah!" moment. You feel a surge of excitement and energy when you've stumbled on an important truth that was previously obscure.

As increasing lines of evidence converge on a conclusion, scientists grow more confident that it is right. Darwin wrote about this feeling as he collected support for evolution from about 1837 until *On the Origin of Species* went to press in 1858. In addition, when many independent lines of evidence support a scientific idea, the burden of proof changes. Before Darwin published his impressive case, the burden was on proving the idea that evolution had occurred. But once Darwin had done so, the burden shifted to those who doubted his conclusion—and the ensuing 160 years of observations and discoveries have piled up immense heaps of supporting evidence. Likewise, disputing a single detail is not enough to disprove anthropogenic climate change. This, too, is a well-established idea supported by reams of data.

The Threads of Toba

In the case of the Toba catastrophe model, a remarkable consilience seems to put the pieces of the puzzle together. We saw in chapter 1 how groups of scientists working with unrelated sets of data on everything from volcanic ash samples to deep-sea plankton found the first evidence for the Toba eruption and its global scale.

A similar thing happened with the human population bottleneck model, as the *Science* journalist Ann Gibbons recounts in her 1993 article "Pleistocene Population Explosions," detailing the consilience that occurred when two groups of scientists found almost the same result independently. Pennsylvania State University anthropologist Henry Harpending was looking at the mitochondrial DNA of many populations of modern humans to find evidence of how long ago they all had a common ancestor. With his former student Alan Rogers (now at the University of Utah), he published theoretical modeling suggesting that there was a population bottleneck some 70,000 years ago, followed by a population explosion about 50,000 years ago.

Harpending had dinner one evening with Mark Stoneking and Linda Vigilant, then also of Penn State but formerly part of the Berkeley group that had recently shown that all modern humans descend from a population that lived in Africa about 100,000 years ago (see chapter 4). They gave Harpending a graph of their mitochondrial data, which showed exactly the kind of distribution that a population would show if it had gone through a bottleneck and then exploded. According to Harpending, "I just about fell out of my chair when Linda and Mark showed me one graph they had."

In August 1993, Harpending, Rogers, Stoneking, and Stephen Sherry of the NIH put all this information about a human population bottleneck into an article they published in *Current Anthropology,* "The Genetic Structure of Ancient Human Populations." It makes a strong case that humans passed through a critical bottleneck and then recovered with a population boom that happened only about 50,000 years ago. This article led to the review by Gibbons, who suggested a connection between this bottleneck and the Toba eruption.

What does this bottleneck mean in the context of the human fossils discussed in the previous chapter? Harpending and others proposed a "Weak Garden of Eden" model for human evolution. According to this model, genetic evidence shows that humans originated in Africa, from which groups migrated at different times. The first of the migrants was *Homo erectus,* which left Africa 1.8 million years ago and reached China and Java by 780,000 years ago. *H. erectus* vanished about 70,000 years ago. Another archaic *H. sapiens* relative, often called *H. heidelbergensis,* appeared in Europe about 600,000 years ago and lasted until about 300,000 years ago. Fossils called "archaic *Homo sapiens*" that are probably about 200,000 and possibly about 300,000 years old have been found in many African and even Middle Eastern localities.

Mitochondrial DNA and fossil skulls suggest both that anatomically modern humans were present in Africa 100,000 years ago (possibly even 300,000 years ago) and that they are the ancestral population of all living humans. They spread out of Africa about 90,000 to 120,000 years ago, as evidenced by finds in the Es-Skhul and Qafzeh Caves in Mount Carmel, Israel, that were sometimes interbedded with deposits bearing Neanderthals. Neanderthals themselves are thought to have separated from the rest of the human family tree about 300,000 years ago, after which they were restricted to the margins of the Ice Age glaciers in

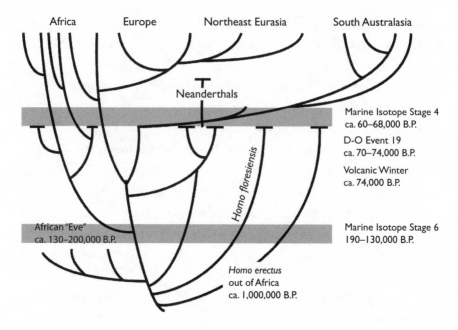

Figure 7.1. Cartoon depicting the human family tree; the genetic bottleneck effect after the Toba eruption, which occurred at the end of Marine Isotope Stage 4 as seen in deep-sea cores, and after Dansgaard-Oeschger (D-O) Event 19 as seen in Greenland ice cores; and the volcanic winter resulting from the Toba ash cloud (*right*). Two human species, *Homo erectus* and *H. floresiensis*, as well as most *H. sapiens* lineages, vanished at this time (*lower right*). Most other *H. sapiens* lineages must also have died out to cause the post-74,000-year bottleneck. After the earth's climate had recovered from post-Toba cooling, the ancestors of all modern humans migrated out of Africa and radiated explosively into different populations in different regions. (Redrawn from several sources)

Europe and the Middle East. They survived until about 30,000 years ago, almost to the end of the last ice age, so they overlapped with modern humans for at least 70,000 years. Genetic evidence even shows that there was a small amount of interbreeding between these two groups, not unexpected for closely related species. DNA analysis of a tiny finger bone and a tooth from a cave in Siberia revealed the existence of another human lineage, the Denisovans, who so far are known only from these remains.

All the foregoing is well-established by this point. The biggest controversy about the origin and migrations of modern humans is over whether modern humans migrated out of Africa and into most of Asia before Toba erupted: if they did, then they apparently were not the ancestors of modern humans, because they predate the bottleneck and vanished from Asia, while post-Toba peoples repopulated the world. Many

anthropologists argue that humans spread to Eurasia only after the eruption, which would be consistent with both archaeological evidence and
the molecular clock dates that have been estimated for the bottleneck.
There are, however, a number of archaeological sites in Asia alleged to
be more than 75,000 years old, though their dating is uncertain and it is
not always clear that they were produced by the same migration of *Homo
sapiens* that produced all modern humans.

While the Penn State geneticists were approaching the problem of
genetic bottlenecks using mitochondrial DNA, Mike Rampino (whom
we met in chapter 1) and Stephen Self of the University of Hawaii were
following up on Craig Chesner's recent discovery that the Toba volcano
had erupted on a massive scale. In two high-profile papers, one in *Nature*
in 1992 and one in *Quaternary Research* in 1993, Rampino and Self made
a strong case that the immense clouds of volcanic ash that Toba shot
into the stratosphere blocked sunlight for years and triggered a gigantic
cooling event. Their data led Gibbons to connect the discoveries about the
human population bottleneck made by Harpending and his colleagues to
that cooling event.

Finally, the anthropologist Stanley Ambrose of the University of
Illinois pulled together all the threads with detailed supporting research.
He was trained in genetics and had a background in studying both human
population genetics and archaeology around the time of the emergence of
H. sapiens. He rigorously put the pieces together in a paper published in
the *Journal of Human Evolution* in 1998 and in a summary he wrote with
Rampino that appeared in 2000.

Bottlenecks Everywhere

Even as geologists were refining the model of how big the Toba eruption
was and how much it affected the climate, and geneticists were looking
more closely at the chronology of genetic changes in humans, other evidence supporting the idea of a bottleneck in the human population came
from unexpected directions, strengthening the consilience of evidence.
In 2004, Shu-Jin Luo and his colleagues sequenced the DNA of all the
living subspecies of tigers. Since tigers, like humans, are land-dwelling,

large-bodied mammals, any volcanic event that has affected humans should have affected them as well. The analysis by Luo and colleagues mapped the interrelationships of tiger subpopulations and determined that their most recent common ancestor lived about 72,000 years ago, when tigers went through their own severe bottleneck.

In 2014, Luo and other colleagues reported finding that genetic bottlenecks were widespread among cat populations in Southeast Asia and that one of the most important occurred about 70,000 years ago. Likewise, several studies have shown that cheetahs also went through a bottleneck at about that time. Finally, there is the case of everyone's favorite animal, the giant panda. Because it feeds almost exclusively on bamboo, its evolution and population size are very sensitive to vegetation changes. When a team of Chinese geneticists sequenced panda DNA in 2012, they found that this species has experienced numerous population bottlenecks and explosions and that a major bottleneck occurred at some point before 50,000 years ago.

Primate research has yielded similar results. A 2006 study by Michael Steiper found that the populations of orangutans in Southeast Asia expanded about 64,000 years ago after a bottleneck. Ryan Hernandez and others showed in 2007 that South Asian macaques went through a population bottleneck around the time of the Toba eruption or shortly thereafter. Even animals on a different continent, far from the volcano, had a population crash. In 2006 and 2011, studies by O. Thalmann and others revealed that the gorillas of Africa experienced a population bottleneck that reduced their numbers to as few as 2,900 mated pairs of one subspecies. Once again, the calculations of genetic divergence suggested that the split from this bottleneck occurred about 77,700 years ago. A 1996 study by T. L. Goldberg showed that chimpanzees experienced a bottleneck dated in the 70,000-year range.

But the strangest of all contributors to this consilience is a bacterium in the gut of more than half of living humans. Known as *Helicobacter pylori*, it is found near the pyloric valve (between the stomach and the small intestine) and has been shown to cause ulcers. As a parasitic infection, it occurs in nearly every human population today, although it produces ulcers only in people whose stress level or stomach condition makes them vulnerable to its effects. When a large group of scientists led by Bodo Linz sequenced samples of *H. pylori* from people all over the earth in 2007, they were able to establish not only how the bacterium first infected our distant African ancestors but also when it spread to populations in Europe. Sure enough, the latter seems to have happened at some point before 58,000 years ago, about the time when human populations exploded in Eurasia after the Toba event.

Almost every organism whose DNA has been sequenced and that had ancestors in Eurasia about 70,000 years ago appears to show the same phenomenon: a population bottleneck about 70,000 years ago, followed by a population expansion about 50,000 years ago. Naturally, most genetic sequencing so far has focused on large, charismatic mammals on which people like to spend research money, such as pandas, tigers, orangutans, chimps, and gorillas. Who knows what surprises lurk in the genomes of the rest of the population of animals that have long lived in Eurasia? Only further investigation will tell.

Testing the Hypothesis

The Toba catastrophe–human bottleneck model hit the world of anthropologists like the shock of a volcanic eruption. As is often the case in science, it naturally generated a storm of controversy. Much has been said and written about the Toba catastrophe hypothesis, and we cannot summarize all of the discussion even in a book like this one. But the salient arguments revolve around the following points.

Was the eruption truly a catastrophe? The volume of ash in the Youngest Toba Tuff (YTT), mentioned in chapter 3, suggests that about 2,800 cubic kilometers (670 mi³) of ejecta erupted, 50 times as much material as in huge recent eruptions such as Tambora or Krakatau. Certainly, ash deposits seem to suggest that it was a huge event. The Toba eruption formed an ash sheet that spread over at least 4 million square kilometers (15 million mi²) and perhaps more than 21 million square kilometers (81 million mi²), or 14 percent, of the earth's surface. The ash is found in nearly every deep-sea sediment core spanning the interval between 60,000 and 80,000 years ago. It is seen across much of the Indian Ocean and is found up to 4 centimeters (1.5 in) thick in the South China Sea, more than 1,800 kilometers (1,100 mi) east of Toba, and in the Arabian Sea, about 4,400 kilometers (2,700 mi) northwest of Toba. On land, it has been found in numerous archaeological sites and once covered almost all of South Asia with a blanket about 15 centimeters (6 in) deep.

Shortly after the publication of the Toba catastrophe model, climatologists such as Alan Robock showed that the volcanic winter following such a massive eruption would have caused decades-long regional cooling of as much as 15°C (27°F) and global cooling of 3° to 5°C (5° to 9°F). In 2002, the volcanologist Clive Oppenheimer argued that this supposition is too extreme and that the amount of global cooling is more likely to have been only 1° to 2°C (2° to 4°F), but in 2009, Robock and colleagues demonstrated that there were mistakes in Oppenheimer's assumptions and the earlier cooling values are realistic.

The eruption that created the YTT produced worldwide cold and dark that seldom let up—the Year without a Summer on steroids. The tree line would have dropped by about 3,000 meters (9,900 ft) as most of the world's higher elevations returned to a glaciated state. Sander van der Kaars and colleagues showed in 2012 that pollen samples from deep-sea cores prove that the Toba eruption caused a major expansion of cold-adapted pine forests about 74,000 years ago. In 2000, Rampino and Ambrose argued that cold-sensitive tropical rain forests would have been nearly destroyed, at least close to the source of the eruption, because of severe drought and disruption of the monsoonal rain pattern. Even grasslands would have been decimated and would have recovered only slowly as the planet warmed after the volcanic winter. In 2009, Ambrose, Martin Williams, and others examined pollen from a number of Indian Ocean sediment cores that show that land plants changed dramatically across South Asia around 70,000 years ago. The same authors also looked at ancient fossil soils preserved beneath and above the Toba ash in India and found abundant evidence of cooling and drying shortly after the YTT eruption.

The crucial question, which if answered affirmatively would establish the link between the eruption and climate, is whether the Toba event coincided with a huge episode of cooling and glaciation known as Greenland Stadial 20 (GS-20), which began about 74,000 years ago and lasted more than 1,000 years. The Toba eruption shows up as concentrations of sulfur dioxide in Greenland ice cores (hence the stadial name), and the chemical signal of the oxygen molecules in the cores allows us to estimate the earth's surface temperature at the time. The most recent studies confirm earlier estimates of the severity of the cooling. A group led by Victor Polynak demonstrated in 2017 that even cave formations (stalactites and stalagmites) in New Mexico record a big drop in temperature about 74,000 years ago, as well as prolonged drought. More important, they found that there was a cooling event in both hemispheres; it was not restricted to the Asian tropics, where it had begun. Polynak and colleagues also confirmed that the cooling caused by the Toba eruption was responsible for the cooling event that followed GS-20, the long interglacial episode when temperature fluctuated between really warm and slightly cold (fig. 7.2). The ice age that thus began about 74,000 years ago did not come to an end until 17,000 years ago, when the earth entered an interglacial period that lasted 10,000 years and resulted in an explosion of human civilization, the birth of history, and the end of the Ice Age megamammals.

The Critics Strike Back

Whenever a new scientific idea is proposed, it is subject to peer review by even the harshest critics. It must withstand the scrutiny of skeptics trying to prove it wrong. The more outrageous the idea, the more closely it will be reviewed and the more stringently it will be critiqued. As is appropriate for such a provocative idea, several groups of scientists have criticized the Toba hypothesis, and their objections have been widely cited. For example, Chris Clarkson and colleagues published a paper in 2012 that documents archaeological sites in southern India whose existence appears to have encompassed the time of the Toba eruption. These scientists say that there was no noticeable change in the artifacts at these

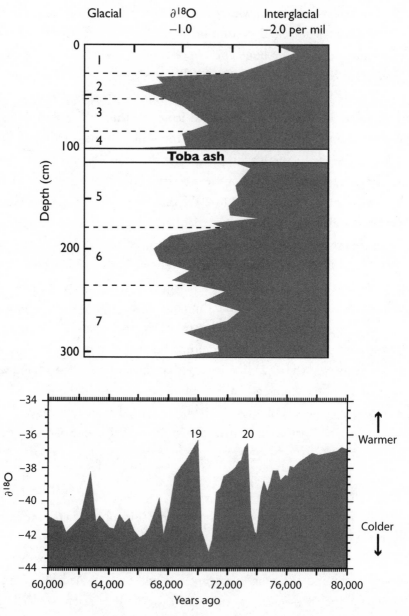

Figure 7.2. Time scale of climate change over the past 100,000 years, showing fluctuations and occurrence of ash from the Toba eruption. *a.* Temperature curve recorded in deep-sea cores taken around the world. The farther the gray area shifts toward the left, the colder the temperature. Numbers on the left are oxygen isotope stages (OIS) 1–7. The top axis is δ¹⁸O, the ratio of oxygen-16 to oxygen-18, which is a function of ocean temperature. The Toba ash coincides with a rapid cooling event between OIS 4 and 5, about 72,000 years ago. *b.* Record from Greenland ice showing warming and cooling cycles as recorded by oxygen isotopes in air bubbles trapped in the ice. The Toba ash coincides with the cooling event that ended Greenland Stadial 20. (Redrawn from several sources)

sites as a consequence of the eruption, which therefore did not disrupt the civilizations that created them. Earlier, in 2007, Michael Haslam and Michael Petraglia argued that an archaeological site in India, known as Jwalapuram locality 3 site, shows outcroppings of YTT but does not show signs of change in the archaeological artifacts located just below and above that tuff, similarly suggesting that the eruption didn't alter the lives of the people who created them. But, as Martin Williams and others argued in 2010, there are problems with the 2007 study. The supposed pre-YTT artifacts suggesting human occupation in India before the eruption do not seem to be in their original positions, so there is no proof that they are older than the YTT. In addition, this archaeological site does not disprove the severity of the eruption or the idea that it might have caused a population crash. The Toba event would have greatly reduced the population of Asia, but there is no reason to think it would have changed the local culture completely, or caused a massive change in the style or abundance of the artifacts found there. Indeed, a few thousand remaining humans might be expected to have clung to their old technologies in their desperate efforts to survive. The archaeological record in this case is inconclusive. Other research suggests that there *were* effects on Asian cultural sites, including a study published by Sacha Jones in 2012 that demonstrates subtle changes in the archaeology of Indian sites that date to around the time of the YTT.

As this book went to press in 2018, Eugene Smith and colleagues reported in *Nature* that two archaeological sites on the coast of South Africa show no obvious change in the artifacts found above and below the level of the Toba eruption. However, a striking sandy layer is seen at the Toba horizon there, which suggests that a dramatic drying and cooling event happened around the time of the eruption. In the absence of any high-resolution environmental data from above and below the ash horizon, it is pointless to speculate about its impact on these coastal sites. And, as in the previous studies, the lack of change in the style or abundance of artifacts doesn't prove the sites' population size or whether a population crash occurred there.

Another line of evidence relevant to Toba's effects comes from lake deposits in eastern Africa that span the time of the eruption. Studies in 2007 by Andrew Cohen and others and by Christopher Scholz and colleagues show that there was severe aridity in eastern Africa around the time of the eruption—but critics of the Toba model minimize these data by claiming the lake sediments are poorly dated. Christine Lane and colleagues argued in 2013 that sediments in Lake Malawi, in southeast Africa, show no evidence of substantial temperature change during the eruption or its aftermath. However, as Richard Roberts and colleagues pointed out that year, findings from this single lake contradict the findings on all the other lakes in the 2007 studies. In addition, Toba's environmental effects might not have been very pronounced in a region as far away as eastern Africa, which also sits right on the equator, where the stability of the tropics could have tamped them down. A 2015 study by Lily Jackson attempted to support the conclusion that the eruption did not affect Lake Malawi, but it did not address the well-documented contemporaneous changes in many other eastern African lakes.

What about the mammalian response to the Toba catastrophe? Nothing on this topic has yet been investigated for African mammals. A 2012 study by Julien Louys looked at fossil mammal assemblages from the Pleistocene in eastern Asia, but the data do not seem to show much convincing evidence of either response or nonresponse to the Toba climate event.

The Toba event and its effects are so controversial that a major symposium was held on the topic, and its proceedings were published in a special volume of *Quaternary International* in 2012. The papers range in opinion from highly supportive to highly critical of the Toba bottleneck hypothesis. But even the critical papers don't provide strong evidence against the YTT ashfall having had a big effect in places such as India or Africa. If the skeptics wished to sink the Toba model, they failed to do so.

In 2017, another startling discovery was reported. In the high-profile journal *Geology*, Jessica Tierney and Paul Zander from the University of Arizona and Peter de Menocal from Lamont-Doherty Earth Observatory described data from a deep-sea core taken from the Gulf of Aden, north of Somalia and south of the Arabian Peninsula. This core contains sediments that accumulated for more than 140,000 years, spanning the past two glacial and interglacial episodes. Using the latest data on genetic divergence as well as data on archaeological sites in the Old World, these scientists showed that anatomically modern humans were restricted to Africa and parts of Israel and the Arabian Peninsula until the arrival of a relatively cool, dry climate episode known as Marine Isotope Stage (MIS) 4, dated at 75,000 years ago. In contrast to what many archaeologists have claimed, there is no good evidence that modern humans had reached most of Asia before then, and surely not by MIS 5, dated between 80,000 and 120,000 years ago, when the climate was milder and wetter. As Tierney, Zander, and de Menocal pointed out, the mitochondrial DNA of women with a distinctive gene group shows that all non-African people alive today share a common ancestor who lived about 75,000 years ago. Similarly, the Y-chromosome DNA of living men with another distinctive gene group shows that non-Africans did not begin to spread into Asia until about the same time. This puts an upper limit of about 75,000 years ago on the exodus of modern humans from Africa and the Near East. The lower limit is established by the founding of the Eurasian genetic groups, which occurred around 50,000 years ago, and the timing of the interbreeding of anatomically modern humans with Neanderthals, about 56,000 years ago. In addition, the conclusion that humans had arrived in Asia by about 58,000 years ago is supported by genetic data showing that the forebears of today's Aborigines finally made it to Australia about 50,000 years ago. Thus, the newest data from both climate and genetics converge on a common answer: anatomically modern humans did not live outside Africa and the Near East until sometime after the Toba eruption—just as the Toba catastrophe model predicts.

Where Do We Stand Now?

As the dust (scholarly, not volcanic) settles, 20 years after Ambrose published his fully developed Toba catastrophe model and 30 years after volcanic ash in ice cores and deep-sea cores was first linked to Toba, it's worth taking stock of what we know and what we don't know.

Two decades of study, criticism, and new data have certainly strengthened the conclusion that the eruption was huge, that it radically changed the climate of much of the earth, and, with the latest dating, that it triggered the thousand-year cooling episode that followed GS-20. Ice cores, deep-sea cores, and deposits found in sites across southern Asia show that huge areas were blanketed with a lot of volcanic ash. Models of climate change based on the amount of ash injected into the stratosphere by Toba leave us little choice but to admit that the climate must have become abruptly cooler and drier over a large area, which is confirmed by the analysis of ancient vegetation. All this evidence is now well established.

Meanwhile, critics of the Toba model have shown only that some lakes in Africa were not as sensitive as others to the climatic effects of the eruption, which also didn't change artifacts in India much—but again, there's no reason to think that tool kits should change just because population size dropped dramatically.

The major question mark is still the dating of the genetic bottleneck. By their very nature, estimates of when a population dropped based on modeling and back-calculating are imprecise, so we cannot point conclusively to the human genome and say the bottleneck happened exactly when the Toba catastrophe took place. Everybody—both supporters and critics of the Toba model—admits this. However, the imprecision of the bottleneck's date is not evidence against the bottleneck's connection to Toba. In addition, there is no other event we know of that might have reduced human population size and diversity so drastically. A study by Per Sjödin and others in 2012 reconfirmed the timing of the bottleneck as approximately coincident with the Toba event, as did the 2017 Tierney, Zander, and de Menocal paper. There is also the consilience of the ages of the genetic bottlenecks—about 74,000 years ago—in so many other organisms, from pandas, tigers, cheetahs, macaques, orangutans, chimps, and gorillas to something as tiny as a bacterium that causes ulcers in people. This remarkable fact suggests that something must have decimated the populations of land animals (probably including humans) within the window of time that includes the Toba eruption.

Although nobody can say with confidence that the Toba catastrophe hypothesis is established and widely accepted, most of the evidence either strongly supports it or does not conclusively contradict it. The fact of a major climate change and the beginning of a thousand-year cold spell around the time of the eruption is clear. This returns us to the original puzzle: Why are humans so low in genetic diversity? What caused the human population to drop to just a few thousand breeding pairs? If a huge event—the Toba eruption and the subsequent climate change—occurred at the time of the genetic bottleneck, isn't that the simplest explanation? We can't prove conclusively that Toba produced the bottleneck, but it's the best explanation we have at the moment. Like most ideas in science, it is tentatively accepted until better data come along.

All this examination of volcano-triggered population crashes raises another question: did such a thing ever occur *before* Toba? Indeed, it turns out that volcanism has been a major cause of mass extinction in the past, including three of the four largest mass extinction events ever documented. We'll look at the evidence in the next chapter.

8

Volcanoes of Doom

Mass extinction is box office, a darling of the popular press, the subject of cover stories and television documentaries, many books, even a rock song. . . . At the end of 1989, the Associated Press designated mass extinction as one of the "10 Top Scientific Advances of the Past Decade." Everybody has weighed in, from the *Economist* to *National Geographic*.

David Raup, *Extinction: Bad Genes or Bad Luck?*, 1991

The "Big Five"

The Toba eruption may have precipitated a population crash that wiped out many human and animal populations about 74,000 years ago. However, volcanism has caused much more severe events in the geological past, not only devastating animal populations but also driving many species to complete extinction. The worst of these events annihilated about 95 percent of all species on earth and almost extinguished animal life altogether.

In 1982, the paleontologists Jack Sepkoski and David Raup of the University of Chicago published an analysis of huge sets of data about fossil species, including dates of the appearance and disappearance of families, genera, and species from the fossil record. They found that five large extinction events stand out from the usual background of extinction events that occur every few million or so years. The Big Five (fig. 8.1) are listed below.

1. The Late Ordovician extinction, 444 million years ago;
2. the Late Devonian extinction, about 360 million years ago;
3. the Permian-Triassic extinction, the biggest of all, 250 million years ago, at the end of the Permian Period;
4. the Triassic-Jurassic extinction, about 200 million years ago; and
5. the Cretaceous-Paleogene extinction, which wiped out the nonbird dinosaurs, about 66 million years ago.

Ever since these five events were singled out as unusual and worthy of further research, there has been no shortage of ideas proposed to explain them. For a long time, scientists tried to blame all five mass extinction events on huge meteorite impacts, major climate changes (both cooling and warming events), and even eruptions of supervolcanoes. Since the early 1980s, however, research has reached some conclusions on the causes of the Big Five extinctions and debunked many of the first wild proposals. Today, more than 30 years after the peak of the frenzy of ideas, the debate continues, but in a much less heated manner.

What has emerged is that there is no common cause for all five of the big mass extinctions. Nor is there a single evolutionary pattern that unites them all. This came as a blow to scientists who had suggested that mass extinctions might be periodic, occurring every 26 million or so years. Among them were Sepkoski and Raup, who, in 1984, published a paper supporting this idea. While that paper was still awaiting publication, a number of astronomers jumped on their bandwagon, proposing earth-impacting objects as causes of this alleged periodicity. They postulated that periodic comet showers, the oscillation of the solar system through the galactic plane, or even the influence of an unknown Planet X or an undetected companion star to the sun named Nemesis could be causes. Other scientists suggested that there was a 26-million-year periodicity in mantle overturn within the earth, triggering pulses of volcanism and global climate change that caused extinctions.

As the biologist Thomas Henry Huxley observed in his presidential address to the British Association in 1870: "The great tragedy of Science—the slaying of a beautiful hypothesis by an ugly fact." No

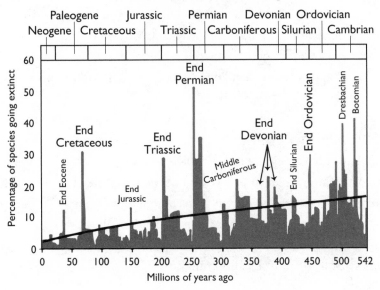

Figure 8.1. The Big Five extinction events plotted as diversity through time. (Redrawn from several sources)

evidence for a Nemesis or Planet X has ever been found, or for comet showers, or for mass extinctions somehow caused by the earth's motion through the galactic plane. Similarly, the mantle periodicity model has been discredited. In fact, the very existence of the 26-million-year extinction cycle has been challenged on statistical grounds: the extinction peaks in Sepkoski and Raup's model are not spaced exactly 26 million years apart, so they don't meet the statistical definition of *periodic*, and some are so small that they don't qualify as statistically meaningful. Much of their original fossil data set was based on species and genera that were not even real to begin with but were instead the result of bad or outdated taxonomy.

Finally, in 1990, the paleontologist Steven Stanley showed why mass extinctions tended to be spaced roughly (but not exactly) 26 million years apart and don't happen more often. After a truly giant mass extinction, so much of life's diversity vanishes that for about 5 to 10 million years, most of the extant species are descendants of the extinction-resistant survivors.

These species tend to be opportunistic, hardy generalists that can live almost anywhere and recover quickly, just as weeds do a few days after you finish gardening. It takes a minimum of 20 million years for highly specialized species that are vulnerable to extreme environmental conditions to evolve and reoccupy the old niches. Thus, a major mass extinction can't happen fewer than 20 to 26 million years after a previous mass extinction.

Now that the debate has cooled and most of the wilder ideas have been dumped into the dustbin of discredited scientific hypotheses, it is clear that only one of the Big Five is related to evidence of the impact of an asteroid or other planetary body. Severe climate change (especially rapid global cooling) seems to have driven two others. But the three remaining of the Big Five are associated with the biggest volcanic eruptions in earth's history. If there is any single event that is a species killer, it seems to be a gigantic volcanic eruption.

The Doom of the Dinosaurs

The mass extinction that is most famous is the one that wiped out the dinosaurs 66 million years ago. Most people know nothing of the species that disappeared during the other mass extinctions, but dinosaurs are hugely popular today. At the peak of the debate over the dinosaur extinction, it even made the May 6, 1985, cover of *Time* magazine and was frequently mentioned in the media.

This topic was barely discussed before 1980. Most ideas about why the dinosaurs died out were speculations with no testable consequences that could allow the scientific community to evaluate them. Some blamed the dinosaurs' demise on cooling or warming climate, their inability to digest the newly evolved flowering plants, mammals that ate their eggs, diseases, and even depression and global weltschmerz. A few blamed it on alien invasions. Many of these ideas were clearly false: for example, flowering plants could not have caused the dinosaurs' extinction, because they evolved 40 million years before the dinosaurs vanished, and they may have spurred the evolution of specialized herbivores such as duck-billed dinosaurs. Likewise, dinosaurs and mammals both appeared about

220 million years ago, and they lived side by side for at least 130 million years. Thus, there is no reason to think that mammals suddenly ate all the dinosaur eggs.

The biggest problem, however, was that explanations for the extinction at the end of the Cretaceous Period, 66 million years ago, were too dino-centric. The event used to be known in shorthand as the K-T (or KT) extinction: the *K* stands for Cretaceous (because *C* was already used for the Carboniferous Period, so instead we use the *K* from the German word for the Cretaceous, *Kreide*, or "chalk") and the *T* for Tertiary Period. This use of the antique term *Tertiary* is now considered obsolete by groups such as the International Commission on Stratigraphy, so today the event is called the Cretaceous-Paleogene (K-P or K-Pg) extinction. It wiped out not only the dinosaurs (although not their bird descendants) but also most of the marine reptiles, including mosasaurs and plesiosaurs; a variety of marine invertebrates, such as the nautilus relatives known as ammonites; and much of the marine plankton. Any explanation that focused on only the dinosaurs missed the global nature of this extinction and failed to show why plants and animals from every level on the food chain also died out. The end of the dinosaurs was really an afterthought. If plankton, ammonites, many other marine invertebrates, and most land plants were wiped out, then it makes sense that the top of the food pyramid would also disappear.

The once quiet, speculative discussion of dinosaur extinction exploded into a frenzy after one crucial, serendipitous discovery. In 1978, the UC Berkeley geologist Walter Alvarez (whom I knew when I was a grad student at Lamont-Doherty Geological Observatory and he was there as a postdoctoral researcher) was studying deep marine strata near the town of Gubbio in the Apennine Mountains of Italy as part of routine geologic mapping completely unrelated to the extinction event. Between the Cretaceous limestones and the Paleocene limestones was a thin clay layer. Alvarez wanted to estimate how much time this clay represented.

His physicist father, Luis Alvarez, suggested checking it for cosmic dust, which falls on the earth in a steady rain from space: a low content would suggest that the clay had accumulated rapidly, while a high content might indicate a long-term accumulation. The best marker of cosmic origin seemed to be a rare platinum-group metal, iridium, which is found in trace amounts in extraterrestrial rocks and in the earth's mantle and core but not its crust. Walt sent his samples to the Berkeley nuclear chemists Frank Asaro and Helen Michel, who found far more iridium than even the longest possible time of accumulation could explain. Luis Alvarez wracked his brain for an explanation, and the only one that made sense was an impact by an asteroid about 10 kilometers (6 mi) in diameter. It was pure serendipity: the Alvarezes were looking to understand how long a geologic event had lasted and instead stumbled upon evidence of something much more dramatic.

After the Alvarez, Alvarez, Asaro, and Michel paper was published in 1980, the iridium anomaly and the impact hypothesis were questioned, especially on the grounds that marine clays are notorious for accumulating all sorts of trace elements. But when high levels of iridium were found not only in other marine strata but also in coal layers that mark the K-P boundary on land, there was little doubt that whatever had deposited it must have been a global event. Since then, an iridium anomaly has been found in more than 100 K-P boundary sequences around the world. Unlike previous speculation about what had killed the dinosaurs, however, this model generated independent predictions that motivated many scientists to look for further evidence. Within a few years, much more evidence had been found, including tiny spherules that appeared to be melt droplets, or crustal material that was liquefied and scattered by the impact; a kind of quartz called stishovite that forms only in shock events; grains of quartz with features showing that they had undergone extreme shock, as in a nuclear blast; angular deposits of debris ejected from an impact crater; and deposits evidently left by a tsunami, which could have been caused by an asteroid, in several places around the Caribbean.

In 1978, Glen Penfield, a petroleum geologist working in Mexico, had identified a likely impact crater called Chicxulub in the northern Yucatán Peninsula, long since filled in and buried deep under the Mexican jungle. Chicxulub was in the right location, at the rim of the Caribbean, and later research dated it at the right age for the K-P boundary. But Penfield was looking for oil, not craters, so impact advocates completely missed his discovery until 1990, when the planetary geologist Alan Hildebrand independently rediscovered the same crater via Penfield's report and his own research. Given all these clues, there seems to be little doubt that some sort of impact occurred at the K-P boundary.

But is this the whole story? Ever since the impact theory hit the scene in 1980, another group of scientists, led by Charles Officer and Charles Drake of Dartmouth, has argued that the huge eruptions that created the area of solidified flood lavas known as the Deccan traps, in western India and Pakistan, could have been responsible for the K-P extinction as well. These eruptions were the second largest in earth history and came not from a single volcanic crater (like those in Hawaii) but from deep rifts or cracks in the crust through which magma from the upper mantle and lower crust poured in huge gushes, rapidly covering more than 10,000 square kilometers (3,900 mi^2) to a maximum thickness of 2,400 meters (7,900 ft). The Réunion mantle plume, or "hot spot," now sitting under its namesake island in the Indian Ocean, was apparently responsible for these eruptions.

Such mantle-derived volcanism yields immense amounts of greenhouse gases, which can cause global warming. It also can create ash clouds that block sunlight and cause global cooling (as happened at the K-P boundary). The Deccan eruptions could have produced high levels of iridium, just as mantle-derived volcanoes still do today, and may have had an early, explosive phase that created something resembling shocked quartz and melt droplets. More important, the peak of the Deccan eruptions was recently dated at 66 million years ago, the age of the K-P boundary, although they started a million years earlier, in the Late Cretaceous. Thus, a prediction based on timing would distinguish the volcanic from the impact hypothesis: the volcanic effects on the environment and on the extinction of organisms should be gradual and prolonged over thousands of years, but an impact would look like a sharp, instantaneous extinction "horizon," with everything dying out at once.

Here's where the controversy over the cause of the K-P extinction gets even more confusing. Some paleontologists argue that dinosaurs were slowly dying out well before the end of the Cretaceous, while others say that this apparently gradual extinction is only an artifact of sampling, and that it disappears when every last scrap of dinosaur bone right up to the last Cretaceous layer is considered. At one time, paleobotanists said the extinction of land plants was gradual, but later they noticed the abundance of fern spores in coal on the K-P boundary and concluded that the disappearance was sudden. One group of micropaleontologists (microfossil specialists) states that the extinction of foraminifera was sudden at the K-P boundary, but another group has long argued that gradual extinctions of these organisms can be seen on both sides of the extinction horizon. Ammonite specialists were once convinced that ammonites had died out gradually, but some have revised their opinion to the opposite conclusion. Thus, the evidence does not support the idea that all the K-P extinctions occurred suddenly and simultaneously at a single point in time. Many were under way several million years before the impact, a fact that could be explained by an atmosphere that was growing more hostile thanks to the accumulated effects of the Deccan eruptions. Even more damning for the impact model, many of the victims (such as marine reptiles and most of the strange oysterlike groups of mollusks at that time) were clearly gone long before the end of the Cretaceous.

On land, the pattern is even more puzzling. The rat-size mammals then in existence mostly crossed the K-P boundary with no obvious effects, except that placentals replaced marsupials in abundance. Even more striking is the lack of change in terrestrial vertebrates, such as turtles, crocodilians, and salamanders. Most of the impact scenarios postulate a long period of cold and darkness caused by clouds of debris and dust, and some suggest there were global wildfires and acid rain as well. Yet turtles or crocodilians could not have hidden from these conditions any more than dinosaurs could. It is true that some of them can hibernate, but they require a long preparation time beforehand, and the impact would have happened without warning. Besides, the Cretaceous greenhouse world was warm all the way to the poles, unlike the earth of today, so they might not have hibernated at all then. The most sensitive creatures should have been salamanders, frogs, and other amphibians, which cannot tolerate acid waters; in fact, their populations are now in serious trouble because of human-induced acid rain. The heavy acid rain postulated by some impact scenarios would have wiped them off the face of the earth had it occurred.

The impact model was dominant at almost every geological society meeting I attended from my early graduate student days in 1980 through the late 1990s. However, judging from talks given at the annual meeting of the Geological Society of America in the past few years, the volcanic mechanism has regained favor. There is no other major mass extinction associated with an extraterrestrial impact, even though during the 1980s some scientists were boldly claiming that rocks from space were responsible for all mass extinctions. In fact, some of the biggest and best-documented impact events, other than the one at the K-P boundary, caused no extinction whatsoever—so the impact model is itself extinct, except for the K-P event. Today, some 38 years after the asteroid impact model was proposed, the media and many scientists outside paleontology think that a rock from space killed the dinosaurs and that that's the final answer to the question of why dinosaurs and the other victims went extinct. But to paleontologists and many geologists, the answer is still "It's complicated."

"The Great Dying"

The dinosaurs may be the most glamorous extinct creatures, and they also get the most attention, but the K-P extinction was not the largest in earth's history. In fact, it may be only the third or fourth biggest. Less known to the general public is the greatest mass extinction of all time. It occurred at the end of the Permian Period, about 250 million years ago, so it is called the Permian-Triassic (P-T) or end-Permian extinction. This event wiped out possibly 83 percent of all marine invertebrate genera and 57 percent of all families. Paleontologists refer to it as "the mother of all mass extinctions," or the Great Dying, since it was far more severe than any other extinction event.

In the marine realm, the extinction was dramatic: 95 percent of marine invertebrate species vanished. It marked the end of the line for five groups of animals that had survived every other Paleozoic extinction event, although some were already on their way out. The horseshoe crab–like trilobites were already down to just two genera, which vanished in the Permian extinction. The two dominant orders of Paleozoic reef-building corals (which had built enormous reefs in the middle Paleozoic) were wiped out. The blastoids, distant relatives of the crinoids (sea lilies and feather stars), had also been slowly declining since their heyday 360 million years ago. None of these four groups was very diverse when the catastrophe came, but the fifth, the single-cell bottom-dwelling shelled amoebas known as fusulinids, had been diverse and hugely abundant in the Permian and covered the seafloor with their shells (which look like rice grains).

Many other marine orders just managed to struggle through. The most abundant types of nautiluslike ammonoids were so badly impacted that only two or three lineages survived, which became the ancestors of a great radiation in the Mesozoic. The major groups of Paleozoic bryozoans, or moss animals, lost 80 percent of their genera, and 98 percent of both the Paleozoic crinoid genera and the marine snail genera disappeared, leaving only a few minor families to repopulate the world in the Mesozoic. The extinction of clams was not nearly as severe: only 59 percent of the genera disappeared. But 96 percent of the genera of the ubiquitous brachiopods, or lamp shells, which had dominated the shelly fauna of the seafloor, vanished. Only four lineages managed to survive into the Mesozoic: a now extinct order of brachiopods known as the spirifers (which hung on until the extinction at the end of the Triassic) and three families that are still alive today. These three were the primitive lingulids, the oil lamp-shaped terebratulids, and the corrugated rhynchonellids.

Extinctions on land were slightly less severe but still the worst the earth had yet seen. At least 70 percent of all creatures disappeared, including most of the lineages of protomammals that had long ruled the Permian, nearly all the reptilian groups, and most of the archaic amphibians, including the giant, flat-bodied, crocodilelike temnospondyls. We can see the severity of this extinction in places such as South Africa's Karoo Desert, where a huge diversity of protomammals in the Late Permian was depleted to just a few genera such as *Lystrosaurus*, plus a few small predatory creatures and a handful of small reptiles, in the Early Triassic.

At one time, scientists blamed the end-Permian catastrophe on the assembly of the Pangaea supercontinent (shallow seas were wiped out as the continental blocks collided with one another), but that had occurred more than 50 million years earlier. Others assigned responsibility for the extinction to the growth of a great ice sheet on the southern supercontinent of Gondwana (consisting of today's South America, Africa, India, Australia, Antarctica, and Madagascar), but that was already in place 100 million years earlier and was gone by the Late Permian. A few scientists claimed that a meteor or a comet hit the earth then (as happened at the end of the Cretaceous), but this idea did not withstand the scrutiny of other scientists: none of the typical signs of impact, such as melt droplets scattered around the earth or large deposits of rare elements such as iridium, has ever been associated with this event.

Instead, the cause of the mother of all mass extinctions seems to have been something even more frightening: the biggest volcanic eruptions in all of earth's history, which created the Siberian traps. In many places in northern Siberia, these ancient lava flows are still visible. Over the last few thousand years of the Permian, almost 4 million cubic kilometers (1 million mi^3) of lava erupted and covered 2 million square kilometers (770,000 mi^2). These eruptions from deep mantle sources would have released enormous volumes of greenhouse gases, especially carbon dioxide and sulfur dioxide.

The addition of these gases to the atmosphere would have triggered a massive super-greenhouse climate. The oceans would have become supersaturated with carbon dioxide, making them hot and acidic and killing nearly everything that lived there. Ocean temperatures are estimated to have reached more than 40°C (104°F), far hotter than even most tropical life can tolerate for very long. The warming and thawing of the seafloor may have released immense quantities of frozen methane—an even more potent greenhouse gas than carbon dioxide—from bottom sediments. In many places, black shales formed, which are typical of deep, stagnant oceans, and other geochemical evidence suggests that ocean waters were depleted of oxygen and may even have been poisoned by hydrogen sulfide. The atmosphere, too, was low on oxygen and full of excess carbon dioxide, so land animals above a certain size nearly all vanished. Only a few smaller lineages of protomammals, reptiles, amphibians, and other land creatures made it through the hellish world of the Late Permian and survived into the aftermath conditions of the Early Triassic.

In short, although the impact at the end of the Cretaceous has received all the mass extinction publicity—to the extent that it is the first thing most people think about when they hear "mass extinction"—it turns out that the K-P and the P-T extinctions were also characterized by the two biggest volcanic eruptions in the earth's history, which transformed the planet's climate. Just one extinction event (the K-P) is associated with an impact, and that may have been only the coup de grâce delivered to a world already irrevocably poisoned by the Deccan eruptions. The biggest extinction, however, was caused by volcanism, not an impact.

The Third Eruption

Have there been any other major mass extinctions associated with huge volcanic eruptions? Yes, another of the Big Five (fig. 8.1) is probably eruption-related: the Triassic-Jurassic extinction, which happened about 200 million years ago. The end-Triassic, or T-J, event eliminated 48 percent of the marine genera, making it only slightly less severe than three others of the Big Five. This "event" was actually a composite of several extinctions spanning more than 17 million years over the last two stages of the Triassic, the Carnian and the Norian (the last part of the Norian used to be considered a separate stage, the Rhaetian). Upward of 90 percent of the clam species and 80 percent of the brachiopod species—including the spirifers, which dominated the mid-Paleozoic and survived the end-Permian event—that were alive in the Late Triassic went extinct by the Jurassic. The abundant nautiluslike ammonoids were nearly wiped out, with just a handful of genera going through a bottleneck (as they did at the end of the Permian) before radiating profusely in the Jurassic. The hardy conodonts—jawless, eel-like fish first known from tiny tooth-like fossils—had survived all the Paleozoic extinction events but finally gave up the ghost. There were significant extinctions among marine snails and sea lilies as well. Reef communities were affected, too, and the only marine reptiles that survived were the fishlike ichthyosaurs. Land verte-brates were hit very hard, although this happened in two pulses, one each at the end of the Carnian and the Norian. In fact, in clearing the landscape of so many archaic amphibians, nondinosaurian reptiles, and protomammals, these extinctions readied the world for the rapid evolution of dinosaurs in the Early Jurassic.

Like the Late Ordovician, Late Devonian, and end-Permian events, these extinctions were very protracted. Some, especially among the ammonoids, bivalves, and terrestrial and marine vertebrates, were con-centrated in the Carnian. Others—for instance among the brachiopods and conodonts—were concentrated in the Norian, and still other groups gradually declined throughout the interval. The severe extinction in Triassic reefs, especially in the tropical seaways in the middle of Pangaea, suggests that cooling was a significant factor in the T-J extinction. An abundance of black shales and geochemical anomalies all over the world suggest that major oceanic changes were also important.

For a long time, people searching for the culprit pointed to the Manicouagan crater in Quebec, the site of a major impact that might have happened in the Late Triassic. Unfortunately, recent dating of the crater at 214 million years old places it far from the T-J boundary and even further from the age of any other mass extinction. Shocked quartz and iridium have also been claimed for this boundary, but subsequent scrutiny has shown that their concentrations are so small as to make their relation to the extinction unlikely.

So what did happen at the end of the Triassic? Although not as much work has been done on the T-J extinction as on more glamorous events, detailed geochemical analysis of soil horizons has shown that the cli-mate rapidly changed at that point. Most recent research has focused on the huge eruptions of the Central Atlantic magmatic province (CAMP) lavas, which occurred when the Atlantic basin ripped apart as Pangaea broke up. The CAMP is a classic example of what geologists call a large igneous province, and it is the biggest ever discovered. It was created by at least four huge pulses of lava totaling more than 3 million cubic kilometers (720,000 mi³) that erupted over about 600,000 years and blanketed around 11 million square kilometers (4.2 million mi²). Its volume wasn't as large as that of the Siberian eruptions, but it covered a much wider area than any other eruption ever documented.

Volcanoes and Extinction

Three of the biggest mass extinctions in earth's history. Three of the biggest eruptions in earth's history, all known to have caused huge amounts of climate change. This is no chance correlation: these eruptions best explain three of the earth's worst extinction crises. But, as stated before, there is no evidence that volcanism had anything to do with the other two of the Big Five extinctions, the Ordovician and the Devonian.

Nonetheless, as the evidence for gigantic mantle volcanism at the P-T, T-J, and K-P boundaries accumulated in the 1980s and 1990s, several scientists promoted the idea that volcanism had caused most mass extinctions in the planet's history. Mike Rampino and Richard Stothers made a connection in 1988, and the French volcanologist Vincent Courtillot even wrote a book about it in 1999. Once again, these authors argued that the spacing of large volcanic events was both periodic (a concept that had been debunked by 1999) and correlated to extinctions. These claims required stretching the data beyond permissible limits. For example, the extinctions in the Eocene and Oligocene (which I have spent much of my career working on) were blamed on eruptions that occurred around that time in Ethiopia. But when you look closer at Courtillot's plot (fig. 8.2), the dates of the extinction peaks appear to be about 31 million years ago, long after the extinction events at 33 and 37 million years ago—and during a period with no major episodes of extinction in the oceans or on land. Plus, many of the other extinctions on this plot are not real. There were no significant mass extinctions in the middle Miocene, when the Columbia River flood basalts erupted in what are now eastern Washington and Oregon, for instance. In fact, only three of the points in fig. 8.2 are actual extinction events that have been correlated with large eruptions.

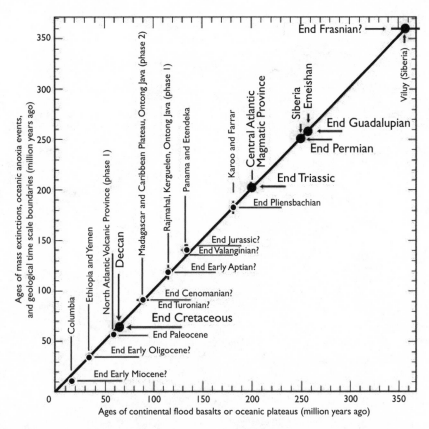

Figure 8.2. Vincent Courtillot's plot of extinction events and volcanism in the geologic past, promoting the idea that there is a tight correlation between the two. (Modified from V. Courtillot, *Evolutionary Catastrophes: The Science of Mass Extinction*, Cambridge University Press, 1999)

Let's not get carried away, as too many scientists have done in the past. There is a lot of interesting data out there, but not all apparent connections are real. And as we have seen over and over, extinction events are complex things, often with multiple causes. Any one-size-fits-all model or explanation is almost certainly wrong.

Future Shocks

Despite our strongly felt kinship and oneness with nature, all the evidence suggests that nature doesn't care one whit about us. Tornadoes, hurricanes, floods, earthquakes and volcanic eruptions happen without the slightest consideration for human inhabitants.

Alan Lightman, "Our Lonely Home in Nature," *New York Times*, May 2, 2014

Could It Happen Again?

We can look at the cataclysmic eruptions that wiped out 95 percent of life on earth at the end of the Permian, or the similarly huge eruptions that helped trigger extinctions at the end of the Triassic and at the end of the reign of the dinosaurs, and ask ourselves: Could it happen again? Could another eruption cause a population crash around the globe, as Toba may have done 74,000 years ago? And, of more importance to readers in the United States, how likely is an eruption in North America that would devastate life in this country?

Certainly some huge North American volcanoes have erupted in the recent past. The Cascade Range, in coastal Oregon, Washington, and Northern California, is made entirely of explosive volcanoes. The most recent to erupt was Mount Saint Helens in Washington in 1980, which is known to have been active since prehistoric times. It erupted in 1480, as Native American legends record, and again for 57 years beginning around

1790, as the British captain George Vancouver, one of the first Europeans to explore the region, witnessed in 1792 while sailing up the West Coast.

In March 1980, swarms of earthquakes preceded the most recent large eruption of Mount Saint Helens. They were followed by lots of steam venting from the crater, while the north flank of the mountain bulged outward. Most residents and tourists were evacuated from the "red zone" close to the volcano, but geologists continued to monitor it from what they thought was a safe distance, and some campers and photographers remained nearby as well. The owner of the Mount Saint Helens Lodge on Spirit Lake, Harry Truman (no relation to the former president), refused to leave his home, and he amused reporters for weeks with his colorful chatter and defiance of the mountain.

Then, early on the morning of May 18, 1980, all hell broke loose. First an earthquake of magnitude 5.1 shook the region, and then the entire north flank broke away from the peak and formed a gigantic debris avalanche, the largest in recorded history, which covered about 600 square kilometers (230 mi^2) north of the mountain. A blast of volcanic gases rushed not only upward but also horizontally, to the north, which endangered and even killed people in that direction who had thought they were far enough away to be safe. The cloud of gases and the moving debris were so powerful that they knocked down all the trees in the area in the same direction, as if they had been combed flat. Geologists who later modeled the gas cloud used the dynamics of jet engines to understand its flow patterns. In some cases, the gases swirled backward in turbulent eddies, as evidenced by downed trees whose tops pointed toward the mountain. The US Geological Survey volcanologist David Johnston, who had been monitoring the activity from a base called Coldwater II, 10 kilometers (6 mi) north of the peak, radioed his last words to his headquarters in Vancouver, Washington—"Vancouver, Vancouver, this is it!"—and died instantly, asphyxiated and burned by the volcanic gases. Most of the nearby landscape vanished under the pyroclastic flows, including Spirit Lake, and with it Harry Truman.

The eruption produced many other dramatic phenomena. Gigantic lahars, or hot volcanic mudflows, roared down the Toutle and Cowlitz

River valleys, gathering up huge logjams that destroyed bridges and carrying away cars, logging trucks, and houses. The lahars transported about 3 million cubic meters (100 million ft³) of material more than 27 kilometers (17 mi) to the Columbia River. The ash from the eruption plume rose into the stratosphere for more than nine hours (fig. 9.1), forming a dense blanket across eastern Washington and parts of Idaho and Montana, and even reaching as far as Edmonton, Alberta. When it was all over, 57 people had died and 250 homes, 47 bridges, 24 kilometers (15 mi) of railways, and 298 kilometers (185 mi) of roads were destroyed.

Visitors to Mount Saint Helens today can see the dacite dome that has been slowly growing (accompanied by small ash and steam eruptions) for the past 38 years. It is most easily viewed through the gap on the north side of the horseshoe-shaped crater. You can travel to the Mount Saint Helens National Volcanic Monument, whose visitor center on Johnston Ridge (named after the volcanologist who died in the event) explains the eruption. On your way over the roads to and from the visitor center, you'll drive through kilometers of moonscape where broken trees lie where they were blasted down. Only a small amount of vegetation and wildlife has come back.

The southernmost volcano in the Cascades was the next most recent to erupt, roughly a century ago. Just south of Mount Shasta, Mount Lassen has erupted many times over the past 5 million years, covering much of northwest California with volcanic debris. After 27,000 years of dormancy, it rumbled back to life on May 30, 1914. Area residents witnessed the beginning of a full year of steam explosions, more than 180 in total, which opened a new crater in Lassen's dacite dome. About a year later, on May 14, 1915, incandescent blobs of lava could be seen from more than 32 kilometers (20 mi) away. Five days later, the main eruption began. In its first phase, the crater blew its top and lava spilled 300 meters (1,000 ft) down the side. Then melting snows formed lahars that rolled 11 kilometers (7 mi) down river valleys, wiping out six houses but not killing anyone. The final, climactic eruption came on May 22, when a column of ash rose 12 kilometers (7.5 mi) into the stratosphere. Pyroclastic flows spread across 7.7 square kilometers (3 mi²) of the landscape, especially

Figure 9.1. The 1980 eruption of Mount Saint Helens, Washington. *a.* The eruption at its peak. *b.* Mount Saint Helens today, showing a small steam eruption from the dacite dome in the enormous crater, with its north rim blown away. (Both courtesy US Geological Survey)

through the river valleys that the lahars had just devastated; many new lahars were also produced. Ash and pumice from the stratospheric dust cloud blew as far east as Winnemucca, Nevada, more than 320 kilometers (200 mi) away. After the paroxysmal blast of May 22, 1915, Lassen settled down. Only steam eruptions were reported over the next two years, ending around June 1917. The volcano now appears to be dormant, although every once in a while steam erupts from its flanks. However, geologic hazards remain in the area: rain and snow could form lahars, which in turn could mobilize old pyroclastics. Should Lassen grow restless again, of course, there is also the possibility of future pyroclastic flows and explosive eruptions.

The deadliest of the Cascade volcanoes has not erupted since there has been a written historical record of the region, but its terrifying past literally looms over the Seattle-Tacoma area in the form of Mount Rainier. People in Seattle celebrate clear weather, when they can see Rainier, and they've named a local beer after it, but many do not realize that the mountain is a dormant volcano with a long history of destruction. Its earliest dated eruptions happened about 2.9 million to about 840,000 years ago and produced a proto-Rainier. Most of the present volcano, built through successive eruptions, is more than 500,000 years old. It was heavily carved by glaciers in the late Ice Age, about 20,000 years ago, and these young glacial valleys have not been obliterated by a new eruption, which demonstrates that Rainier has not completely blown its top since that time. Nevertheless, there have been small eruptions in parts of the volcano, the most recent one occurring from 1820 to 1854.

Rainier's largest geologic event was an eruption about 5,600 years ago that produced the Osceola mudflows. These lahars removed almost two to three cubic kilometers (0.5 to 0.7 mi^3) of rock from the summit, which they spread across the nearby river valleys, covering about 550 square kilometers (212 mi^2) northeast of the mountain. Almost all the towns since built in those valleys—including Tacoma, Auburn, Kent, Enumclaw, Sumner, and Puyallup—sit right on top of the old lahar deposits. The summit of Mount Rainier also collapsed during the eruption of the Osceola mudflows, lowering it by about 7 kilometers (4 mi).

The avalanche of hot volcanic mud, rock, and trees traveled at more than 6 meters (20 ft) per second, destroying everything in its path. It even entered the legends of the local Native Americans.

About 500 years ago, a smaller lahar from Mount Rainier, known as the Electron mudflow, swept down the Puyallup River valley, knocking down trees 3 meters (10 ft) wide; the town of Orting is built on its deposits. The volcano has not had a significant eruption since, but many geologists worry that one of the greatest hazards it presents is not a gigantic eruption (which hasn't happened in a long time) but a smaller one that would unleash similar lahars on the huge populations of the Seattle-Tacoma suburbs.

These recent eruptions are tiny, however, compared to the biggest events in the Cascade Mountains. By far the largest was the catastrophic explosion of Mount Mazama, which became Crater Lake. Mazama began to grow about 400,000 years ago and eventually became one of the tallest stratovolcanoes in the chain. Flows of andesite lava (which are unusual, since most Cascade volcanoes erupt dacite or rhyolite) poured down the north and southwest slopes about 50,000 years ago, raising the volcano's height to about 3,400 meters (11,000 ft). However, the magmas then changed from andesite to dacite, becoming more and more silica-rich, which made them more viscous and sticky. About 40,000 years ago, numerous dacitic eruptions created a series of domes (much like the one now in the crater of Mount Saint Helens) on the mountain. These were all destroyed in a later series of eruptions and lahars that also left large landslide deposits on all sides of Mount Mazama. The next eruptions, of rhyodacite magma, with a composition between dacite and rhyolite, happened 25,000 to 30,000 years ago. These thick, pasty flows emerged from vents on the northwest flank, forming Redcloud Cliff and the dome above Steel Bay. Then the volcano went dormant for 20,000 years. During that time, large glaciers carved into it, forming channels that are still visible on the lower slopes.

The final phase began 7,677 years ago with an intermittent, centuries-long series of huge eruptions of rhyodacite that generated giant pyroclastic flows all across the region. Then, about 6,800 years ago, a cataclysmic explosion blew the top off Mount Mazama and sent a column of volcanic ash 1.6 kilometers (1 mi) wide up to 16 kilometers (10 mi) into the stratosphere at twice the speed of sound. This huge column collapsed upon itself, just like the mushroom cloud of an atomic bomb, sending monstrous pyroclastic flows in all directions. Most of them were so hot that they welded ash into the solid rock called tuff. Mazama ash fell on much of the western United States as far as Wyoming and Utah and up into Saskatchewan. The eruptions were so huge (46 to 58 km³, or 11 to 14 mi³, of rock) and so fast that the magma chamber beneath the summit didn't have time to refill, and the entire mountain collapsed into its own empty interior. Rainwater and snowmelt later filled this enormous basin. Volcanologists call the large, ring-shaped depression formed by a collapsed crater a caldera, and thus, technically speaking, Crater Lake should in fact be Caldera Lake (although it is impossible to get people to call it by its correct name). There are a number of other calderas on the east side of the Cascades, notably the huge Newberry Caldera near Bend, Oregon.

Supervolcanoes in America

By far the biggest supervolcano in North America was La Garita, in the San Juan volcanic field of southwestern Colorado. All that remains of it now is a huge caldera, 35 by 75 kilometers (22 by 47 mi), which forms a large, irregular, oblong valley in the San Juan Mountains. Some of the caldera and its ash deposits can be seen in the Wheeler Geologic Area northwest of South Fork, Colorado. In other places, the caldera has been filled in by material from eruptions that formed smaller but younger calderas, such as Creede. When La Garita exploded 27.8 million years ago, it blew more than 5,000 cubic kilometers (1,200 mi³) of material, enough to fill Lake Michigan, across a vast swath of the United States. It deposited a widespread ash layer known as the Fish Canyon Tuff, which can be seen in the Arkansas River canyon 100 kilometers (62 mi) northeast of the caldera, and beneath the Alamosa area, 100 kilometers east of the volcano. This is thought to be one of the largest volcanic deposits in the world and probably once covered much of the Rockies and the Great Plains of the United States. It is calculated that the energy released by La

Garita was 10,000 times more powerful than the largest nuclear explosion that humans have ever generated.

After La Garita, the list of supervolcano eruptions is dominated by events in the Yellowstone Caldera complex. Yellowstone sits over a huge hot spot in the earth's mantle, which provides a constant flow of heat (hence the area's geysers and hot springs) and occasional magma. The biggest eruption occurred about 2.1 million years ago, when the mountain that became the Island Park Caldera—the largest in the complex, reaching into Idaho from the west side of the present national park—expelled more than 2,500 cubic kilometers (600 mi^3) of gases and ash, creating the Huckleberry Ridge Tuff. This ash sheet was so enormous that it covered the entire western United States and reached as far as Texas to the south and Missouri to the east. Yellowstone produced two supervolcano eruptions from its Helse volcanic field, also in eastern Idaho, as well. The first, 6 million years ago, spewed more than 1,500 cubic kilometers (360 mi^3) of debris around the region, creating the Blacktail Tuff. The second, dated at 4.5 million years ago, released more than 1,800 cubic kilometers (430 mi^3) of debris, forming the Kilgore Tuff. The smallest supervolcanic eruption in Yellowstone happened just 640,000 years ago (late in the ice ages) and spread more than 1,000 cubic kilometers (240 mi^3) of material over most of the western United States, reaching as far as Louisiana and Arkansas and forming the Lava Creek Tuff (fig. 9.2). This young eruption was the source of the ring-shaped caldera now in the center of Yellowstone National Park. Dozens more eruptions have been documented for the Yellowstone mantle hot spot in Idaho and Wyoming, but only these four qualify as supervolcanic.

The list of supervolcano eruptions around the world includes two in the Taupo Volcanic Zone of New Zealand's North Island and three in the Andes (one each in Chile, Argentina, and Bolivia). Just slightly smaller was the eruption that produced Long Valley Caldera, near Mammoth Mountain Ski Area and the town of Bishop in eastern California's Owens Valley. Dated at 758,900 years ago, this was one of the largest eruptions in North American prehistory. It released 600 cubic kilometers (140 mi^3) of ash, which spread across almost all of western North America,

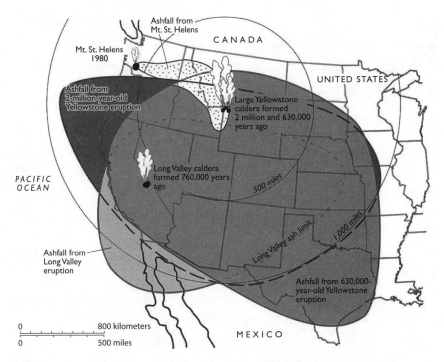

Figure 9.2. Distribution of giant Ice Age volcanic ash clouds, including two from Yellowstone; the Long Valley eruption of Bishop Tuff from Mammoth Mountain, California; and the relatively small eruption of Mount Saint Helens. (Redrawn from several sources)

including parts of Canada and Mexico, and as far east as Nebraska and Kansas. Huge ignimbrites as hot as 820°C (1,500°F) covered the area around Long Valley, forming the thick Bishop Tuff near the vent, which filled the 3-kilometer- (2-mi-) deep caldera nearly to the rim after the volcano collapsed into its own magma chamber. Long Valley Caldera has been dormant since that time, but the many hot springs in the area show that magma is still not far beneath the surface there. In the 1980s, there was much seismicity in the region (including several magnitude 6 quakes), causing geologists to worry that another eruption was on the way. However, that activity died down in the following decades, so the warnings have been scaled back.

North America has long been home to volcanoes that erupt catastrophically and blow out huge volumes of debris, which have often blanketed half the area of the current United States. None of these eruptions except La Garita was as big as Toba, but each would have dramatically impacted life and the environment for years to come. If the death and destruction caused by Mount Saint Helens seem intolerable, just imagine how much worse an eruption of a Yellowstone caldera or Long Valley Caldera would be to American population and infrastructure. The effects would dwarf those of any other natural disaster we have ever experienced, including giant hurricanes such as Katrina and Harvey and great earthquakes such as the one that destroyed San Francisco in 1906. The only reason that we don't realize the seriousness of eruptions or take more precautions against them is that on human time scales, huge eruptions are rare.

A Perspective on Catastrophes

The terrifying power and destructiveness of volcanoes get our attention, at least when eruptions are happening. But we need to keep these disasters in perspective as well. As Mike Rampino pointed out in a 2002 article on supereruptions as threats to civilization on earthlike planets, they tend to happen on average about every 50,000 years somewhere on earth. The really huge flood-lava eruptions tied to major mass extinctions are even more rare, with only three occurring in the past 250 million years.

By contrast, asteroid impacts happen on average once every 100,000 years, only half as often as supereruptions. And despite all the excitement about impacts causing extinction or climate change, only one has been linked to a major mass extinction (the Cretaceous-Paleogene event; see chapter 8). In fact, two of the Big Five mass extinctions have no connection to either an impact or volcanism. In recent years, geologists have also learned that there are no impacts at other mass extinction horizons—and many impacts, such as those that formed craters in Chesapeake Bay and what is now eastern Siberia about 35.5 million years ago, caused no extinction whatsoever.

Although big events such as eruptions, extraterrestrial impacts, and earthquakes attract attention, it turns out that more frequent and mundane events are much more deadly to humans. Many people are deathly afraid of earthquakes, primarily because they shatter our sense that the ground beneath us is terra firma and because they are still completely unpredictable. Yet earthquakes in the lower 48 United States have killed fewer people (only 357 in the past century, or about 3.6 per year) than any other natural disaster, including tornadoes, hurricanes, floods, and blizzards. In fact, your chance of dying in an earthquake in the United States is much less than your chance of being struck by lightning or bitten by a venomous snake. Thus, the fear of quakes in the United States (except in a few areas, such as California) is almost entirely unjustified: they are extremely rare, and even when they do occur, US building codes are so strict that quakes rarely kill many people.

The story is different in underdeveloped countries, especially in Asia and the Middle East. In those areas, not only are earthquakes frequent, but most structures are poorly made of stacked bricks or stones and can become major killers in an earthquake. As they say in seismology, earthquakes don't kill people—buildings kill people. After a big quake in Turkey, Iran, Armenia, Nepal, or China, local people, who often lack access to relatively safe materials such as wood or steel, rebuild with unreinforced brick and masonry just as they did before, resetting the death trap that could kill more people in the next quake.

Even landslides, tornadoes, and hurricanes, which are deadlier than volcanoes or earthquakes, on average together make up less than 13 percent of the fatalities from natural disasters in any given year in the United States. How about floods? Despite the huge floods that afflicted many areas of the United States during the hurricane season of 2017, they account for only about 14 percent of annual US deaths due to natural disasters. Thus, the natural events that terrify people the most—earthquakes, volcanoes, floods, hurricanes, and tornadoes—are much less deadly, in terms of average casualties per year, than they seem.

So what *are* the deadliest natural disasters in the United States? Most people are surprised to learn that they are the most common and mundane occurrences of all: cold and hot weather, especially in the form of heat waves and blizzards. In most parts of the country, people cope with one or both of these kinds of events every year, so they take them for granted. But blizzards and other winter weather phenomena are responsible for about 18 percent of all natural hazard–related deaths in the United States each year. The biggest killer of all is heat waves, which account for almost 20 percent of US natural-disaster deaths annually. People tend not to notice how deadly hot and cold weather are, because snow and heat waves happen every year and aren't fast-moving, dramatic catastrophes like tornadoes, hurricanes, floods, or earthquakes. But pay attention to the news and you will see that record heat waves kill dozens to hundreds every year (mostly elderly people living in facilities without air conditioning), while blizzards trap and freeze people on a regular basis every winter.

Of course, the real dangers to our existence are even more subtle than weather. Natural disasters are scary and deadly, but between 1970 and 2004, they killed only 20,000 Americans. By contrast, in the same period in the United States there were over 30 times more deaths from heart disease: 652,000 (most of them smoking-related). There were also 600,000 deaths from cancer, more than a third of them related to smoking. The rest of the major medical killers, too, are far more deadly than anything nature throws at us: those 34 years in the United States also saw 143,000 deaths from strokes, 130,000 from chronic respiratory diseases such as bronchitis and pneumonia, and about 117,000 from accidents (mostly car accidents). Nature may seem dangerous to us, but our cigarettes, fatty foods, and cars are far more likely to kill us.

We have examined many different ways that humans and other animals can die through nature's wrath, from the large view of mass extinctions spaced out over tens of millions of years through the deadly volcanic events of the past 28 million years to the better-known stories of huge volcanic eruptions that have killed people in historic times. We've also learned that volcanic events have caused far more mass extinction than the much more well-known scenarios such as asteroid impacts. But among all such events in the past 66 million years, the Toba eruption stands out.

Not only was it one of the largest eruptions since the end of the Age of Dinosaurs, but it was the biggest to occur in the 7 million years since humans split from the other apes. Even though the case is not yet conclusive, it is beyond dispute that some gigantic event changed global climate around the time that Toba erupted, about 74,000 years ago. It is also clear that there was a giant human population crash and bottleneck at the same time, followed by a population explosion that later gave rise to most of the peoples of Eurasia, the Americas, and Australia—and that nearly all the other large mammals whose DNA has been sequenced suffered similar bottlenecks at the same time, around 74,000 years ago. All these events seem to be closely connected in time, and if Toba didn't cause the bottleneck in human (and other) population, then what did?

Ultimately, the lesson of Toba is that all humans on earth (except a few of the still-surviving ancient African lineages) are extremely closely related, and all of us descend from a common ancestor who left Africa less than 70,000 years ago. The differences between what people call "races" are extremely recent and biologically meaningless: any two populations of chimpanzees are more distantly related than all the human "races" put together. We are indeed all one "race," we are all brothers and sisters—and, as the singer and songwriter Roy Zimmerman put it, we all have a birth certificate from Africa.

ACKNOWLEDGMENTS

The idea for this book came to me while I was teaching about volcanoes in my geology classes. I always end with Toba and its possible effect on humans. The students never fail to be amazed, and eventually I realized that no one had yet told the story with the proper level of detail or with good explanations of not only the volcanology but also the convergence of molecular biology and anthropology that pulled everything together.

I thank my agent, John Thornton, for getting this book to my publisher; Carolyn Gleason, Laura Harger, Matt Litts, and Jody Billert at Smithsonian Books for their hard work on this project; and Juliana Froggatt for her copyediting. I thank my anthropologist colleagues Greg Laden and Pat Shipman for double-checking my coverage of their profession, T. Ryan Gregory and Norman Johnson for reviewing my molecular biology, and Briana Pobiner, Alan B. Rose, and Martin Williams for reviewing the whole book. I thank my son Erik Prothero for drawing the originals of many of the line images in Adobe Illustrator.

Finally, I thank my supportive family for allowing me time to work on this book: my amazing sons, Erik, Zachary, and Gabriel, and my wonderful wife, Teresa.

INDEX

Page numbers in italics indicate illustrations.